MW00908245

BUILDING EUROPEAN Spatial Data Infrastructures

SECOND EDITION

Ian Masser

ESRI Press
REDLANDS, CALIFORNIA

ESRI Press, 380 New York Street, Redlands, California 92373-8100

 First edition 2007. Second edition 2010.

14 13 12 11 10   1 2 3 4 5 6 7 8 9 10

Printed in the United States of America

Library of Congress Cataloging-in-Publication Data
Masser, Ian.
Building European spatial data infrastructures / Ian Masser.— 2nd ed.
p. cm.
Includes bibliographical references and index.
ISBN 978-1-58948-266-1 (pbk. : alk. paper) 1. Geographic information systems—Europe. 2. Information storage and retrieval systems—Geography. I. Title.
G70.215.E85M37 2010
910.285—dc22   2010014340

Ask for ESRI Press titles at your local bookstore or order by calling 800-447-9778, or shop online at www.esri.com/esripress. Outside the United States, contact your local ESRI distributor or shop online at www.eurospanbookstore.com/ESRI.

ESRI Press titles are distributed to the trade by the following:

In North America:
Ingram Publisher Services
Toll-free telephone: 800-648-3104
Toll-free fax: 800-838-1149
E-mail: customerservice@ingrampublisherservices.com

In the United Kingdom, Europe, Middle East and Africa, Asia, and Australia:
Eurospan Group
3 Henrietta Street
London WC2E 8LU
United Kingdom
Telephone: 44(0) 1767 604972
Fax: 44(0) 1767 601640
E-mail: eurospan@turpin-distribution.com

Table of Contents

Foreword

In February 2006 I was invited to give a presentation at a meeting in Gavle, Sweden, where European GIS experts and politicians came together to discuss INSPIRE. At this meeting the real impact of the INSPIRE Directive and how important it would be to educate EU citizens and politicians about GIS, its applications, and spatial data infrastructures became clear. Many of the attendees agreed that the future of INSPIRE depends in large part on the human factor: understanding the major principles (and especially the possibilities), exchanging experiences and best practices, and sharing data. This sounds easy, but everyday experience reminds us that the hurdles can be considerable.

The European Union is expanding rapidly, and this expansion is creating overpopulation, environmental pollution, traffic congestion, security risks, depletion of natural resources, shortage of drinking water, shrinking biodiversity, and many other problems. GIS can help the EU and its member countries tackle these challenges by helping them make better decisions for a cleaner, safer, and more prosperous Europe. Isolated GIS projects will not do this, but a network of connected regional, national, and pan-European projects will. And that is what INSPIRE is about. It creates the underlying rules for exchanging data and services across boundaries.

I am very pleased that Dr. Ian Masser was enthusiastic when we asked him to write a book explaining INSPIRE to nontechnicians. With his comprehensive knowledge of spatial data infrastructures and his instrumental role in the inception and evolution of INSPIRE, he was the perfect author for such a book.

Jack Dangermond
President, ESRI

Preface

The publication of the first edition of this book coincided with the formal approval of the groundbreaking INSPIRE Directive by the European Parliament and the Council of Ministers in March 2007. Since that time an ambitious programme has been undertaken by the European Commission to create the technical rules that are required so that the member states can successfully implement the provisions of this directive. By the end of 2009 most of these measures were in place or at an advanced stage of development, and member states were also taking steps to transpose the INSPIRE Directive itself into their own legislation. From 2010 the implementation of the directive will be increasingly in their hands. Given these circumstances it was felt that a second edition of this book was required which includes a description of these innovative activities. It has also provided an opportunity to update the discussion of other developments with respect to spatial data infrastructures in Europe.

Acknowledgments

This book is the outcome of discussions between Frank Holsmuller, Regional Marketing Manager, EMEA, ESRI Europe; and Judy Hawkins from ESRI Press regarding the need for a book that would explain current spatial data infrastructure (SDI) developments in Europe, including the pathbreaking INSPIRE initiative to create an infrastructure for spatial information in the community, to a nontechnical audience. I am particularly grateful to Frank for his helpful comments on earlier drafts of this book and for his enthusiastic support for the project. I would also like to thank the multitalented team at ESRI Press for their active involvement in the creation of this book.

The completion of this project would not have been possible without the assistance of lots of other people who are working in this field. I would particularly like to express my appreciation to the following people who have helped me in various ways to complete this work: Alessandro Annoni, Dimitris Assimakopoulos, Mike Batty, T. O. Chan, Max Craglia, Hans Dufourmont, Mike Gould, Iain Greenway, Hugo de Groof, Dietmar Gruenreich, Jordi Guimet, Randy Johnson, Werner Kuhn, Leona Lees, Jenny Lisapaly, Bastiaan van Loenen, Suzanne McLaughlin, Tom Modderkolk, Leen Murre, Martin Peersmann, Guenther Pichler, Gabor Remetey, Jens Riecken, Daniele Rizzi, Francois Salge, Robert Tomas, Saulius Urbanas, Danny Vandenbroucke, Marc Vanderhaegen, Antti Vertanen, Robin Waters, and Ruzena Zimova.

Last but not least, I must thank my wife, Suzy, for her continuing support in completing this project.

Taddington, England
2010

1 The Many Uses of GIS

In April 2006, ESRI President Jack Dangermond gave a presentation at the European Union (EU) Interparliamentary Conference on the INSPIRE (Infrastructure for Spatial Information in Europe) Directive in Gavle, Sweden. The presentation was entitled, "How will we use spatial information in the future and how can a common EU infrastructure for such information contribute to development?" It described a spectrum of geographic information system (GIS) applications across many sectors of society and illustrated the extent to which GIS is an application-driven technology continuously finding new realms that can benefit from the power of spatial information.

LOCAL GOVERNMENT

Local government agencies are the largest users of GIS. Most local government services have a spatial dimension: urban and regional planning, transportation, parks and recreation services, housing, local economic development, property tax collection (box 1.1). GIS applications have been developed to increase the efficiency and effectiveness of local government services at all but the smallest local authorities. Planners of education and health facilities make extensive use of the technology to identify areas of need and to locate facilities, and police rely on it for crime mapping and emergency management.

INCREASING EFFICIENCY AND COLLABORATION

- Urban planning
- Services provision
- Recreation facilities
- Property tax collection
- Local economic development

MODERNISING WORKFLOWS AND PROVIDING ACCESS

- Electricity networks
- Gas supplies
- Water distribution
- Telecommunications
- Cable television

Box 1.1

Managing local government and utilities.

Figure 1.1

GIS can be used to reroute utility cables and sewer pipelines and plan new traffic patterns (Rotterdam, the Netherlands). Courtesy of Frank Holsmuller.

UTILITIES

Because of their enormous investment in fixed infrastructure networks, the electricity, gas, telephone, and water companies were pioneers in the development of large-scale automated mapping and facility management (AM/FM) GIS applications (box 1.1, figure 1.1). As their interests are primarily commercial, utility companies track and maximise returns on existing assets and strive to achieve high service standards. They also seek out new business opportunities to generate revenue for shareholders. These objectives create an ever-increasing demand for accurate, up-to-date information on the state of the utility's infrastructure, the types and frequency of service problems, and the changing tastes of customers. Utility companies also use GIS to link their databases with other geographical features such as road networks, census information, and topographic data.

TRANSPORTATION AND LOGISTICS

The planning and management of road, rail, sea, and air transportation networks require a substantial investment in geographic information technologies. The objectives include improving traffic flows, lowering operational costs, saving energy, and ensuring safety. GIS is indispensable for tracking vehicles over complex networks, and a wide range of geographic information tools are used for in-vehicle satellite navigation, fleet management systems, logistic scheduling and routing, and intelligent vehicle highway systems (box 1.2).

BUSINESS

In a constantly changing marketplace, successful businesses must strive to reduce costs, improve profit margins, and maintain their competitive edge to retain customers. Business applications of GIS can be found in financial services, insurance, manufacturing, and distribution (box 1.2). The real estate industry has a strong interest in geographic information on individual properties and the general land market for financial investment and property development purposes. Retail companies also need geographic information to complement their customer databases and help them find the best locations for new stores and offices.

LOWERING COSTS, SAVING ENERGY, AND IMPROVING TRAFFIC FLOWS

- Road networks
- Railway systems
- Air routing
- Sea lanes

PROVIDING THE GEOGRAPHIC ADVANTAGE

- Financial and insurance services
- Manufacturing goods distribution
- Retail site selection
- Real estate services
- Property investment

PROVIDING VISUALISATION AND UNDERSTANDING

- Water quality
- Pollution levels
- Environmental degradation
- Coastal zone management
- Desertification

Box 1.2

Managing transport, business, and natural resources.

NATURAL RESOURCES

One of the greatest challenges facing the world is the need for more effective management of natural resources, with the goals of maintaining the health of the environment, conserving biodiversity, and facilitating sustainable and equitable development (box 1.2, figure 1.2). Playing a major role in meeting this challenge, GIS is used in applications dealing with salinity, declining water quality, coastal zone erosion, desertification, and environmental degradation. Logging, oil, and gas companies were also among the first private-sector businesses to experiment with GIS for managing inventories.

ENVIRONMENTAL HAZARDS

GIS is a valuable tool in raising public awareness and improving understanding of environmental hazards such as floods, earthquakes, hurricanes, and tsunamis as well as broad issues such as global warming and climatic change (box 1.3). In playing an important role in minimising environmental risks, GIS helps save lives and reduce damage to property. GIS applications have also been developed for monitoring river water quality, environmental pollution, acid rain deposition, and thermal discharge from power stations.

Figure 1.2

Increasing residential development and the growing number of winter sports facilities creates problems for environmental management throughout Alpine regions such as the Austrian Tyrol. GIS can be used as a tool for environmental planning and resource management. Photo by the author.

Figure 1.3

Cameras are increasingly being installed in urban areas to ensure the safety of citizens, and GIS can be used to select the best location and help with the monitoring itself (Rotterdam, the Netherlands). Courtesy of Frank Holsmuller.

SAVING LIVES AND PROPERTY
- Environmental monitoring
- Emergency management
- Natural environmental hazards
- Global warming
- Climatic change

CREATING A SAFER SOCIETY
- Modelling risks
- Tracking diseases
- Facility security
- Vulnerability analysis
- Buffer zone protection

CREATING THE NEXT GENERATION OF USERS
- Promoting spatial literacy
- Raising geographic awareness
- Developing skills
- Increasing understanding

Box 1.3
Minimising environmental risks, managing security, and supporting education.

SECURITY

GIS applications are helping to create a safer society by modelling risks, identifying vulnerable areas, and creating buffer zones to protect citizens from terrorist attacks and industrial waste hazards (box 1.3, figure 1.3). They also provide the critical infrastructure required for effective emergency management and for modelling the spread of diseases such as avian influenza in birds, foot and mouth disease in farm animals, and severe acute respiratory syndrome (SARS) in humans.

EDUCATION

In the last decade, the number of GIS users throughout the world increased from thousands to millions. In the process, the GIS community has grown from the original core of surveying and mapping specialists to include a wide variety of professionals and amateurs who are working with spatial information on many different levels. This rapid expansion has created a need for educational programmes that explain basic geographic concepts and encourage new users to think spatially and increase their awareness and understanding of geography (box 1.3).

CONCLUSIONS

GIS applications span a very broad range, as can be seen in the above examples, and new uses are emerging all the time. However, it must be recognised that the full potential of GIS can be realised only if the necessary spatial data infrastructures (SDIs) are in place at the local, national, and transnational levels. The relationships between GIS and SDIs will be discussed in the next chapter.

How GIS Technology Works and Why We Need SDIs

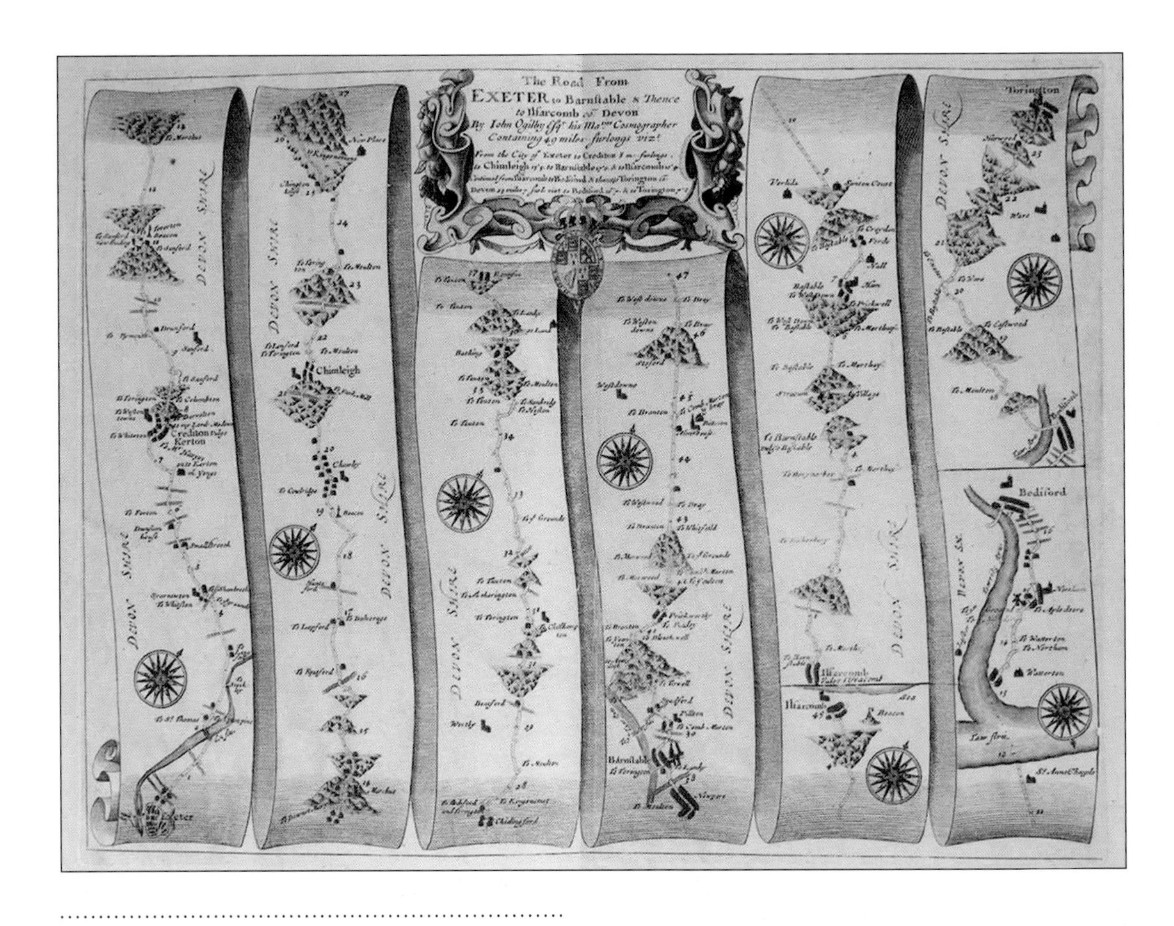

Figure 2.1

John Ogilby's 1675 route map from Exeter to Ilfracombe in England.

LOCATION, LOCATION, LOCATION

Geographic information identifies or describes locations on the surface of the Earth. In the past, such information took the form of paper maps. Now, geographic information can be stored digitally, allowing it to be processed by computers, and this has enabled the creation of many new applications.

Geographic information can consist of addresses, market research data, census data, health data, data on environment and natural resources, descriptions of transportation and utility networks, information on flows of goods, cadastral and land registration information, as well as data obtained by remote sensing from satellites in space.

The simplest way of presenting geographical information is the map. The oldest known maps date back thousands of years. Maps have been preserved on Babylonian clay tablets from about 2300 BC. Modern mapping dates from the Age of Discovery in the fifteenth and sixteenth centuries. This was also a time when more and more people were travelling around by themselves and required basic information in the form of road maps, specifically the kind produced by John Ogilby in England (figure 2.1). The science of cartography grew up around the production of maps.

Maps are widely used because everything that happens, happens somewhere, and knowing where something happens can be strategically important. Consequently, military considerations played an important role in the development of cartography. They were instrumental in the establishment of the Survey of India in 1767 and the creation of the Ordnance Survey of Great Britain in 1791.

Geographic information can be used to answer the following general questions:

- *Where is it?*—Location of a point on the Earth's surface
- *How far is it?*—Distance between two points on the Earth's surface
- *Is it near somewhere?*—Relationship between two points on the Earth's surface
- *In which district does it lie?*—Information about administrative areas
- *How can I get there?*—Information such as the quality of transportation routes and bus and train timetables

Three geometric concepts underlie these questions: **points,** objects whose location is specified by a set of coordinates; **lines,** connected coordinate points forming features such as rivers and roads; and **polygons,** areas bounded by connected lines such as administrative districts and areas of homogeneous land use or soil type (figure 2.2).

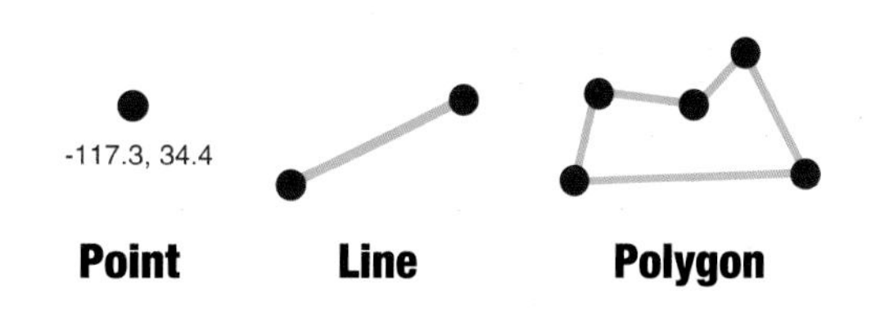

Figure 2.2

Points, lines, and polygons.

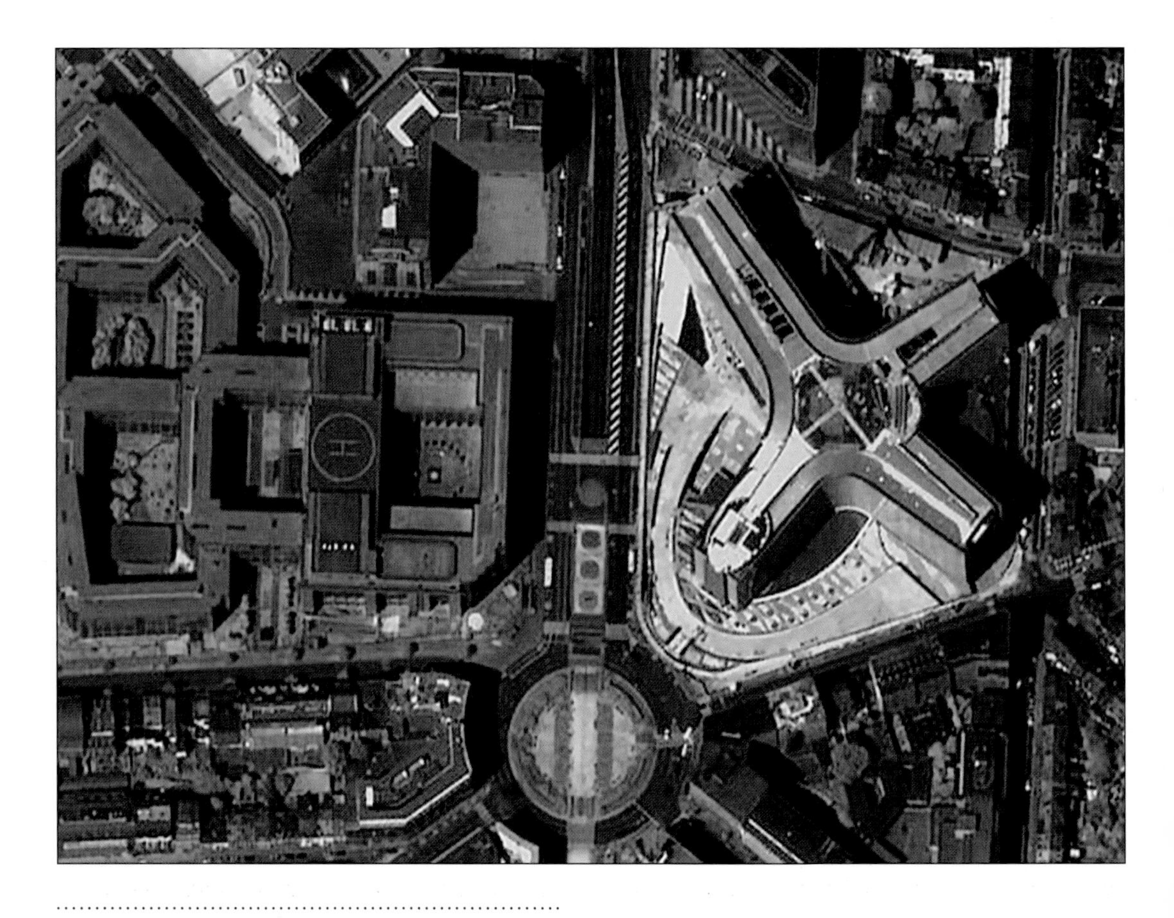

Figure 2.3

High-resolution image of the European Commission Berlaymont Building in Brussels. Courtesy of Digital Globe.

To answer the last of the questions on the previous page, attribute data about the nature and frequency of buses and trains is needed. Attribute data is typically statistical data related to points, lines, or polygons. Examples of attribute data include population census statistics, various kinds of survey data, as well as information related to individual addresses such as property valuation, vehicle registration details, and personal health records.

GEOGRAPHIC INFORMATION SYSTEMS

Geographic information systems (GIS) can be defined as computer systems for capturing, managing, integrating, manipulating, analysing, and displaying data that is spatially referenced to the Earth. They were first developed during the 1960s, but it was not until the mid-1980s that computers capable of handling the considerable amounts of data used in these systems came into being. In 1987 the British Government Committee of Enquiry chaired by Lord Chorley hailed the advent of GIS as "the biggest step forward in the handling of geographic information since the invention of the map" (Department of the Environment 1987 [paragraph 1.7]). Developments since that time have massively increased computer storage and handling capabilities, and the Internet has made it possible to transfer large amounts of geographic information from one place to another.

A detailed examination of the tasks that can be accomplished with this relatively new technology makes it easy to see just how powerful GIS really is.

Capturing. Capturing is the encoding of data into digital form, which enables it to be read by a computer. It includes map digitising, direct recording by electronic survey instruments, and the encoding of text and attributes of all kinds. It also includes the millions of images obtained by means of Earth observation technology. For example, figure 2.3 shows a high-resolution image of the European Commission Berlaymont Building in Brussels taken by the QuickBird satellite from 450 kilometres above the Earth's surface.

Managing. Managing is the creation of databases (collections of data) organised according to a conceptual schema. This is usually handled with the help of a database management system (DBMS), which is software used to organise the information. This software typically contains routines for data input, verification, storage, retrieval, and combination.

Integrating. GIS makes it possible to link datasets and to merge and combine different data for the same location (bus and train information, for example). This makes GIS an analytical tool that is fundamentally different from a conventional paper map.

Manipulating. GIS also makes it possible to manipulate large volumes of data for the same area easily and quickly. This capability gives GIS great potential for generating new products and services which add value to existing spatial data.

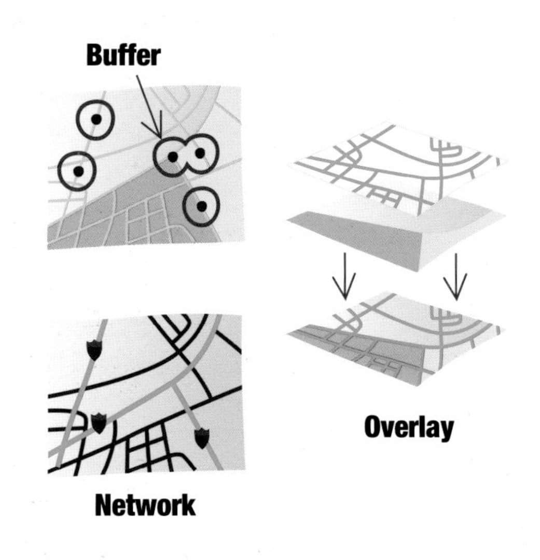

Figure 2.4

Buffer, network, and overlay.

Analysing. The basic analytical capabilities of GIS consist of overlays, buffers, and networks (figure 2.4). An overlay is the superimposition of two or more datasets that are registered to a common coordinate system. Buffers are regions of specified width around a point, line, or polygon. Networks are geometric or logical arrangements of nodes and interconnecting lines. GIS technology also facilitates the development of methods for exploring more complex relationships through mathematical models.

Displaying. Visualisation tools that can be used to display data held in a GIS include not only conventional maps but also graphs and three-dimensional models of surfaces and buildings (figure 2.5).

The benefits of GIS consist of quick and easy access to large volumes of data, the ability to select detail by area or theme, the possibility of linking one dataset with another, and the opportunity to analyse the spatial characteristics of an area as well as new kinds of output tailored to meet particular needs (Longley et al. 2005).

Despite these advantages, the Chorley Committee also concluded 20 years ago that the availability of GIS technologies was "a necessary, though not sufficient condition for the take up of geographic information systems to increase rapidly" (Department of the Environment 1987 [paragraph 1.22]). In other words, political and institutional barriers that currently restrict the use of GIS must be overcome. These include compatibility problems stemming from differences in definitions and formats as well as restrictions on access and the availability of data.

These barriers currently limit the effective utilisation and commercial exploitation of GIS technology, impeding job creation and economic development in the process. They also block opportunities for increasing the effectiveness and transparency of governments.

Figure 2.5

Three-dimensional visualisation from the Virtual London project. Courtesy of Michael Batty, Centre for Advanced Spatial Analysis (CASA), University College London.

To remove the political and institutional barriers blocking the widespread use of GIS, governments need to do the following:

- Ensure that data coverage is comprehensive, the same definitions and formats are used, and the timing of updates is consistent
- Promote interoperability between different datasets and different systems
- Reduce or eliminate restrictions on access and availability while protecting the intellectual property rights (IPR) of data owners and the privacy of data providers
- Disseminate information about what data is available and from what sources

From the early 1990s onward, governments all over the world have felt the need to create spatial data infrastructures. This need has been intensified recently by new technologies such as Global Positioning System (GPS) devices (which give individuals the means of calculating their coordinate references at different locations) (figure 2.6), satellite navigation systems for cars, and a new generation of mobile phone services that can also display map-based information. In addition, new Web-based geographic information services such as Google Earth make it possible for users to view different parts of the world at the click of a mouse. These developments mean that the majority of people, either knowingly or unknowingly, are now users of geographic information.

Figure 2.6

A member of a natural-resources agency uses a GPS device to assist with the monitoring of river levels in the northwest United States. Courtesy of Trimble Navigation New Zealand Ltd.

SPATIAL DATA INFRASTRUCTURES

There are clear parallels between SDIs and other forms of infrastructure. An infrastructure typically comprises the basic facilities, services, and installations needed for the effective functioning of a community or society. Infrastructures include transportation and communications systems, water and electricity services, as well as public institutions such as hospitals, schools, post offices, and prisons.

A spatial data infrastructure (sometimes called geographic information strategy, geospatial data infrastructure, or geoinformation infrastructure) is "the means to assemble geographic information that describes the arrangement and attributes of features and phenomena on the Earth. The infrastructure includes the materials, technology, and people necessary to acquire, process, and distribute such information to meet a wide variety of needs" (National Research Council 1993, 16).

The overriding objective of SDIs is to facilitate access to geographic information assets held by a wide range of stakeholders in both the public and the private sectors in a nation or a region with a view to maximising overall usage. This objective requires coordinated action by governments.

SDIs must also be user driven, as their primary purpose is to support decision making for many different purposes. SDI implementation involves a wide range of activities. These include not only technical matters such as data, technologies, standards, and delivery mechanisms but also institutional matters related to organisational responsibilities, overall national information policies, and availability of financial and human resources.

The SDI phenomenon. Since the term "spatial data infrastructure" was first used in 1991, about half of the more than 200 countries in the world have embarked on some form of an SDI initiative (Crompvoets et al. 2004). Given these circumstances, the term "SDI phenomenon" seems to be a reasonable description of what has happened in this field over the last 20 years. The original leaders in this field were mainly relatively wealthy countries such as Australia, Canada, the Netherlands, Portugal, and the United States, but SDIs are now being developed in all parts of the world. There are considerable differences among countries in terms of both the approach and the content of these initiatives (Masser 2005).

SDIs are under construction at both the national and the subnational levels of government and also at the supranational level. Their primary objectives are typically to promote economic development, stimulate better government, and foster environmental sustainability at all these levels. The notion of better government can be interpreted in several different ways. In rapidly developing countries such as Malaysia, it means better strategic planning and resource development. Planning, in the sense of a better state of readiness to deal with emergencies brought about by natural hazards, was also an important driving force in the establishment of the Japanese national SDI after the 1995 Kobe earthquake. In Portugal, on the other hand, the National Geographic Information System has played an important part in modernising central, regional, and local administration.

The most famous national spatial data infrastructure (NSDI) is the one that was set up in the United States by an executive order from President Clinton on April 11, 1994 (box 2.1). This directive set forth the main tasks to be carried out and defined time limits for each of the initial stages of the NSDI. It strengthened the powers of interagency coordination of the Federal Geographic Data Committee (FGDC), whose membership includes representatives from all the major federal departments with an interest in geographic information and the collection and management of such information. The executive order also required the creation of a national digital geospatial data framework of the most frequently used datasets, the establishment of a national geospatial data clearinghouse to increase user awareness of what data is available, and the facilitation of access to this data. The FGDC Clearinghouse has been one of the most obvious SDI success stories. The FGDC Clearinghouse Registry, for example, lists nearly 500 registered nodes within its network from the United States and other countries. These facilities have been augmented since 2002 by the creation of the Geospatial One-Stop portal to support the E-Government Initiative (figure 2.7).

Box 2.1

Executive order signed by President Clinton in 1994 establishing the U.S. National Spatial Data Infrastructure.

PRESIDENTIAL DOCUMENT

Executive Order 12906 of April 11, 1994

Coordinating Geographic Data Acquisition and Access: The National Spatial Data Infrastructure

Geographic information is critical to promote economic development, improve our stewardship of natural resources, and protect the environment. Modern technology now permits improved acquisition, distribution, and utilization of geographic (or geospatial) data and mapping. The National Performance Review has recommended that the executive branch develop, in cooperation with state, local, and tribal governments, and the private sector, a coordinated National Spatial Data Infrastructure to support public and private sector applications of geospatial data in such areas as transportation, community development, agriculture, emergency response, environmental management, and information technology.

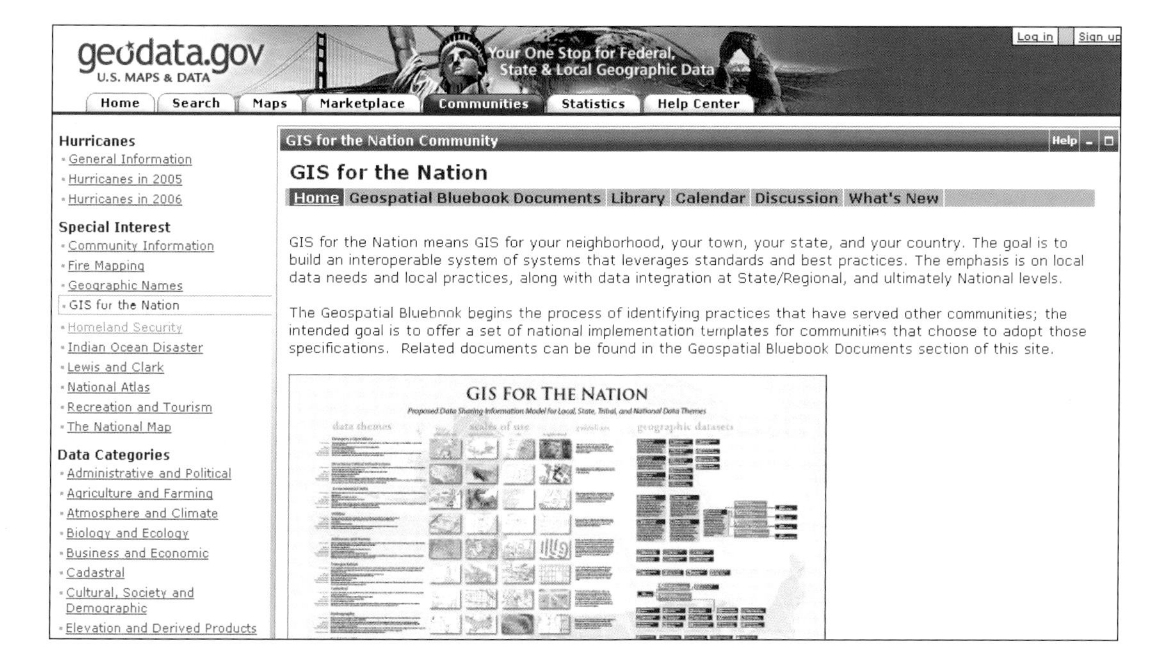

Figure 2.7

Geospatial One-Stop portal search page.

www.geodata.gov.

In contrast to the U.S. NSDI, the lead agency for the Canadian Geospatial Data Infrastructure (CGDI), GeoConnections, has always been a cooperative organization that seeks to involve all the stakeholders from different levels of government, the private sector, and academia, among others (figure 2.8). These interests are reflected in the composition of the GeoConnections Management Board and also in the membership of its constituent nodes such as the Policy Advisory Network. GeoConnections sees itself as a catalyst for successful implementation. There is also a strong industry connection between the CGDI and the Geomatics Industry Association of Canada.

Australia presents another model of SDI governance. The precursor of its Australia New Zealand Land Information Council (ANZLIC) was set up in 1986 as a result of an agreement between the Australian prime minister and the heads of the state governments who wanted to coordinate the collection and transfer of land-related information between different levels of government. Each of the members of ANZLIC represents a coordinating body within their jurisdiction: the Commonwealth Office for Spatial Data Management, the relevant coordination bodies at the state and territory levels, and Land Information New Zealand. In effect, ANZLIC has been developing some elements of the Australian Spatial Data Infrastructure (ASDI) since 1986.

Alongside developments such as these, a number of international regional and global agencies have been set up to promote capacity building and raise awareness of the need for governments to promote the creation of SDIs. These include bodies such as the European Umbrella Organisation for Geographic Information (EUROGI) and the Global Spatial Data Infrastructure Association (GSDI). The GSDI cookbook (Nebert 2004) has been widely distributed over the Internet and played an important role in capacity-building activities all over the world.

SDI components. The process of SDI development and implementation consists of four main components. As can be clearly seen in the Victorian Spatial Information Strategy (VSIS, Australia) for 2004 to 2007 (Department of Sustainability and Environment, Victoria, Australia 2005), the four components are the institutional arrangements that are required for delivering geographic information, tasks related to the creation and maintenance of fundamental datasets, procedures for making geographic information accessible, and ways of facilitating the development of strategic technology and applications (figure 2.9).

Institutional arrangements include matters related to the overall governance of SDIs as discussed above for the United States, Canada, and Australia, as well as the assignment of responsibilities for the custodianship of fundamental datasets.

Victoria's Vicmap gives some idea of the kind of data that is regarded as fundamental to most applications (box 2.2). It includes geodetic

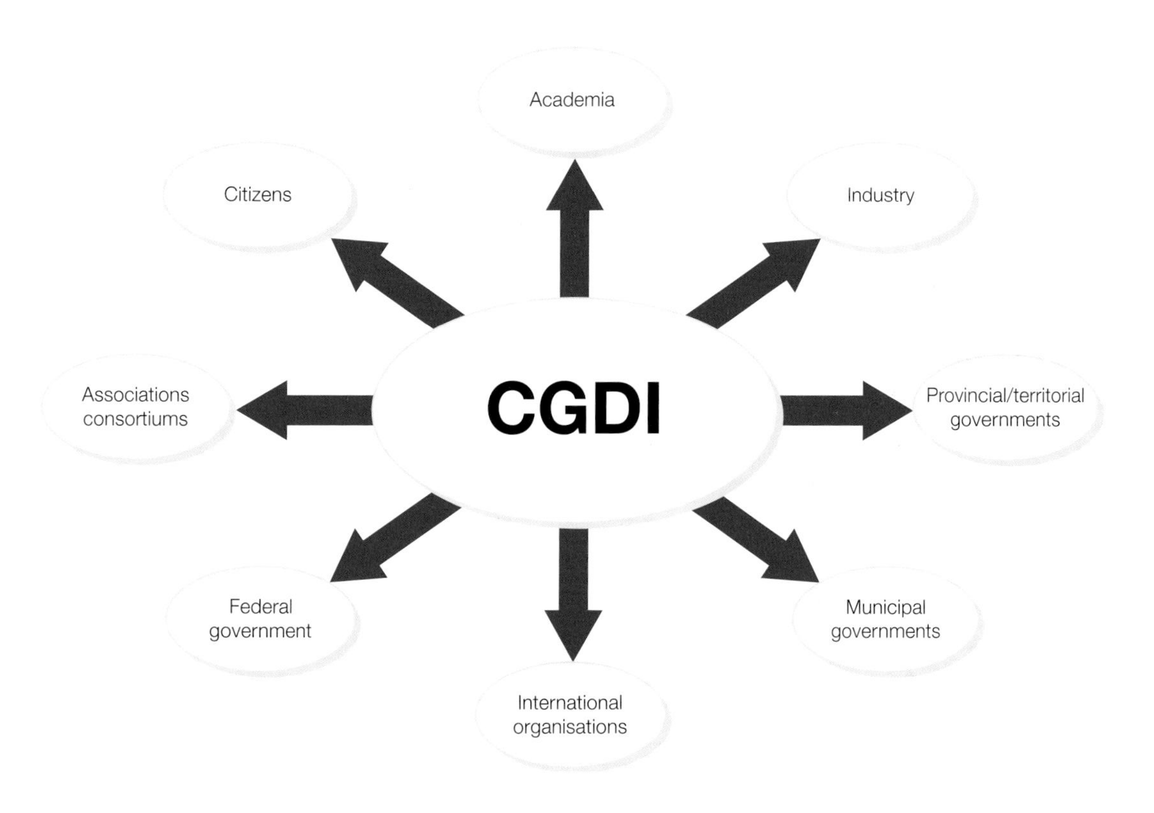

Figure 2.8

Stakeholders in the Canadian Geospatial Data Infrastructure (CGDI). Copyright Natural Resources Canada. Used by permission.

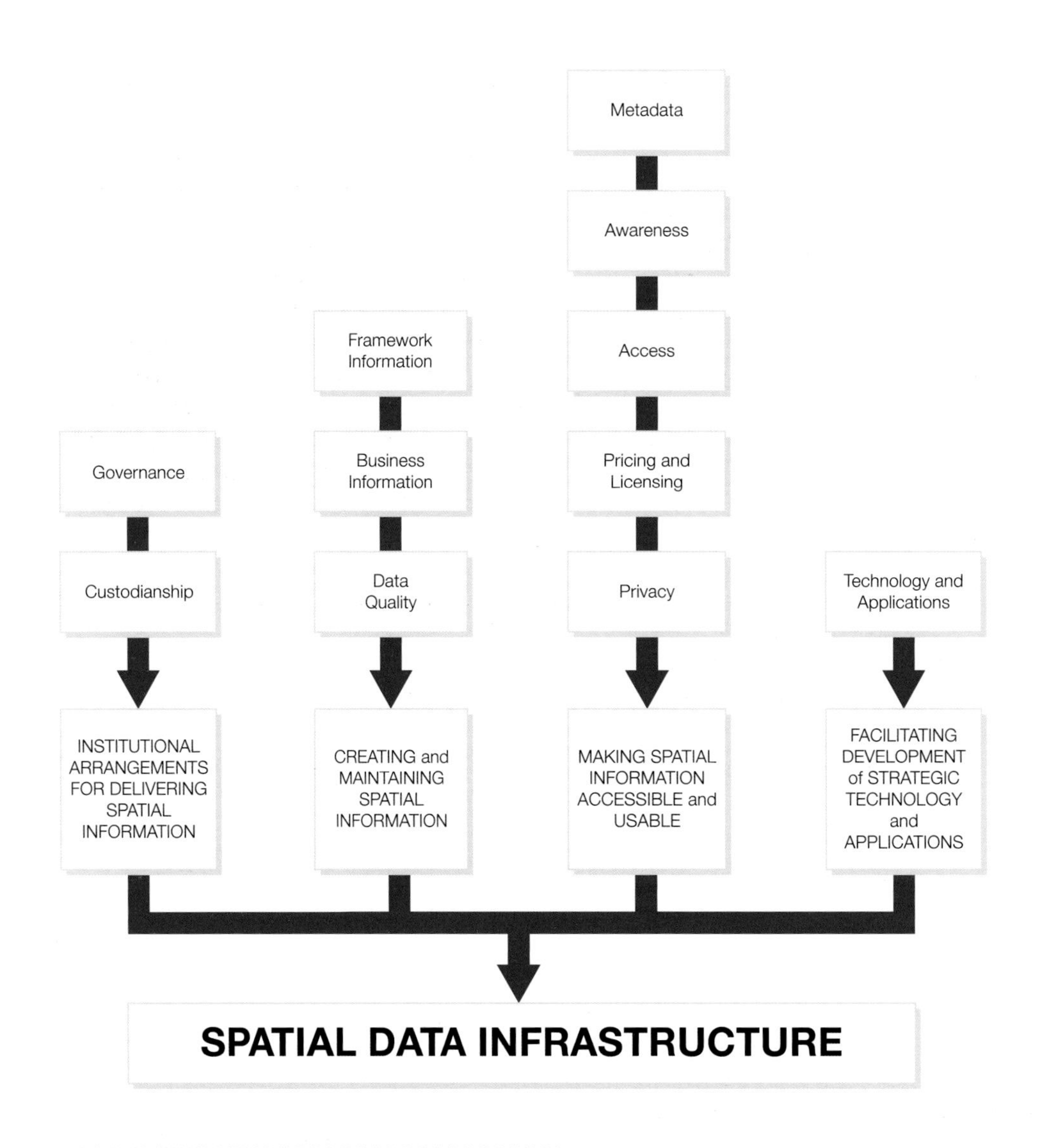

Figure 2.9

Components of Victoria's SDI. Courtesy of Department of Sustainability and Environment.

reference points, elevation and hydrographic data, addresses, administrative boundaries, and property and transport information.

The third component of SDIs is the need for metadata services to increase user awareness of what data is available and to facilitate access to this data. The recent development of spatial portals such as the U.S. Geospatial One-Stop opens up new possibilities for metadata services. Spatial portals can be seen as gateways to geographic information resources which allow users and providers to share content and create consensus. In the process of developing metadata services, it is also necessary to deal with issues such as awareness, access, pricing and licensing, and privacy.

The last component deals with technology and applications. Recent advances in technology as described above have dramatically altered the way in which spatial information is acquired and used. The only constant factor in this demand-driven environment is the need for data that is easy to access and independent of technology or software. This requires that the arrangements for data sharing and exchange be based on the concept of interoperability.

SDI costs and benefits. The costs and benefits associated with SDI development cannot be easily estimated with any precision. Nevertheless, it is clear that the tasks of SDI coordination and governance are relatively inexpensive in relation to the overall expenditure on geographic information, whereas the task of core digital database development is relatively

Box 2.2

Fundamental datasets maintained by the State of Victoria, Australia.

VICMAP DATASETS

Geodesy: More than 100,000 survey marks and a network of GPS stations, collectively known as GPSnet
Property: 2.5 million parcels and properties
Transport: Every road in the state, including forest trails, fire tracks, and property access roads
Address: Administrative boundaries, including locality, electoral, local government, and postcode
Elevation: Slopes, aspects, and contours
Hydrography: All water courses and features
Imagery: Aerial photography as well as Landsat and SPOT satellite imagery

expensive. The U.S. Office of Management and Budget has estimated that U.S. federal agencies alone already spend at least $4 billion annually to collect and manage domestic geospatial data, whereas the costs of supporting the FGDC and its work are less than 1 percent of this amount (Masser 2005, 47).

SDIs have economic, social, and environmental benefits (box 2.3). The most important economic benefit of SDIs is the promotion of economic growth as a result of an expanding market for geographic information products and services both locally and internationally. A PIRA International study conducted in 2000 for the European Commission (PIRA International 2000, 8–9) gives some indication of the value of public-sector information in the tightly constrained European market of that time. The findings showed that over half the economic value of public-sector information in 1999 (€68 billion) came directly from geographic information sources. This is equivalent to about 1 percent of the European gross domestic product. The findings also showed that the economic value of public geographic information resources in the less constrained U.S. market was €750 billion. With the easing of constraints through the development of SDIs, it might be expected that the size of the European market will move toward that of the U.S. market over time. Estimates of the growth of the commercial geographic information market are more readily available and have been on the order of 15 to 20 percent per year in recent years. The impact of an expanding market of this size on job creation is also considerable. Other economic benefits of SDIs include increased efficiency and lower operating costs for both public- and private-sector organisations due to wider access to geographic information and information-based services. These may also be considerable over time.

The most important social benefit of SDIs is the extent to which they create more efficient and more transparent governments at all levels as a result of the increasing availability of authoritative data to policy and decision makers. Another important social benefit stems from the opportunities that SDI data sharing creates for citizens to actively participate in the democratic process. Because they bring together data from many diverse sources, SDIs are also likely to lead to better arrangements for homeland security and more effective systems for emergency planning and response. There are also many operational benefits for social services, public health, education, and public safety from the more effective targeting of areas and groups with special needs.

SDIs can bring many environmental benefits, and they have an important role to play in promoting sustainable development throughout the world. At the national and local levels they provide the data required for effective management and monitoring of natural resources. They are particularly useful in coastal zones because of the extent to which they can integrate maritime and terrestrial data.

ECONOMIC BENEFITS

- An expanding internal market for GI products and services
- Greater competitiveness, more opportunities to export GI products and services
- Increased efficiency of both public- and private-sector organisations
- New opportunities for GI business applications and services
- Improved transport and infrastructure management systems

SOCIAL BENEFITS

- Improved national and local governance
- More opportunities to engage in the democratic process
- More effective homeland security
- Faster emergency response
- Opportunities to target groups and areas with special needs

ENVIRONMENTAL BENEFITS

- Promoting sustainable development
- Better natural-resource monitoring and management
- Improved coastal-zone management

Box 2.3

Benefits of SDIs.

CONCLUSIONS

In order for the enormous potential of GIS to be fully exploited, governments need to ensure that fully compatible and integrated databases are made available without restrictions on reuse of data, thus facilitating access to geographic information assets held by a wide range of stakeholders. Widespread implementation of SDIs will bring substantial economic, social, and environmental benefits.

Existing SDIs in Europe

The development of SDIs has been studied extensively in Europe over the last ten years. This is partly due to the interest of the European Commission in such activities as those expressed initially in the GI 2000 initiative and more recently in the INSPIRE programme (discussed in chapter 4). As a result, the commission has funded a number of important studies in this field. These include the Methods for Access to Data and Metadata in Europe (MADAME) project (Blakemore et al. 1999) and the Geographic Information Network in Europe (GINIE) project (Craglia et al. 2003). Two of the three European Commission Directorates General responsible for the INSPIRE programme also commissioned studies of the state of play of SDI activities in all the European countries from the Spatial Applications Division of the Katholieke Universiteit Leuven (SADL) in Belgium. The study began in 2002, and the reports have been revised and updated every year up to 2007 (http://inspire.jrc.ec.europa.eu/). Another series of state-of-play studies began in 2009. These studies constitute a major resource describing the evolution of SDI development throughout Europe.

The initial findings of the Leuven studies suggested that only a handful of European countries had anything like a full-fledged SDI either planned or in place. Even countries like the Netherlands and Portugal did not meet all of their criteria for a complete SDI as "in neither case are all components of a theoretical SDI in place or even planned" (Van Orshoven et al. 2003, 9). As a result the authors claimed that most of these "NSDI initiatives can therefore better be described as SDI-like or SDI-supporting initiatives."

Since that time considerable progress has been made, prompted to a large extent by the INSPIRE initiative. The authors of these studies have also developed a useful typology of SDIs that is based on the coordination aspects of these initiatives. Matters of coordination are emphasised because "it is obvious coordination is the major success factor for each SDI since coordination is tackled in different ways according to the political and administrative organisation of the country" (Van Orshoven et al. 2003, 9). A basic distinction is made between countries where a national data producer such as a mapping agency has an implicit mandate to set up an SDI and countries where SDI development is being driven by a council of ministries, a GI association, or a partnership of data users. A further distinction is then made between initiatives that do and do not involve users in the case of the former and between those that have a formal mandate and those that do not in the case of the latter (table 3.1). Each of the European Union countries can be classified according to these criteria.

Two thirds of the SDI initiatives were led by national data producers in 2007. This was particularly the case in the central and eastern European countries that became members of the European Union in 2004 and the Nordic countries. All the Nordic countries explicitly included data users in the coordination process, whereas only a minority of the new EU members have involved users. Not all of these SDI initiatives were operational.

SDI STATUS	LED BY NATIONAL DATA PRODUCERS	NOT LED BY NATIONAL DATA PRODUCERS
	Users involved	**Formal mandate**
Operational	Denmark, Finland, Hungary, Iceland, Norway, Portugal, Sweden	Belgium (Flanders), Czech Republic, Germany, Netherlands, Spain, Switzerland
Partially operational	Austria, Greece, Luxembourg, Poland	Ireland, Italy
Not operational	Belgium, Romania	
	Users not involved	**No formal mandate**
Operational	Slovakia, Slovenia, Lithuania	Belgium (Wallonia), United Kingdom
Partially operational	Cyprus, Estonia, Latvia, Liechtenstein	France
Not operational	Bulgaria, Malta, Turkey	

Table 3.1

SDIs in Europe in spring 2007. Source: Vandenbroucke and Janssen. (2008, 40)

FINLAND

The SDI development process started in Finland in the 1980s. Prior to 1996, it was customary to refer to the "shared use of geographic information." The NLS (Maanmittauslaitos) SDI team produced a vision paper on a national geographic infrastructure for Finland in 1996, but it was not until 2001 that the Council of State established the Finnish Council for Geographic Information to oversee SDI activities and INSPIRE process. This was a high-level body with a formal mandate to develop strategies for SDI implementation. Its 17 members included all the key stakeholders in central and local governments and the private sector. The council published the National Geographic Information Strategy for 2005–2010 in autumn 2004, setting out the principles, objectives, and measures needed for the development of an SDI in Finland. This strategy, based largely on INSPIRE principles, was expected to foster more efficient and greater diversity of usage of the available databases and facilitate the emergence of new services as a result of better access to information. The implementation of the 2005–2010 strategy was partly delayed because of the preparation of the INSPIRE Directive. In 2009 a new law concerning the implementation of SDI (INSPIRE) was enacted. According to the law, the responsibility is shared between the organisations managing the core datasets and the National Land Survey (NLS) as an implementation support organisation. The NLS is also responsible for the discovery service and other centralised network services as well as some reporting activities. More than 100 datasets have been identified as INSPIRE datasets. A new structure for open SDI–INSPIRE cooperation (INSPIRE-network) was set up in spring 2009, and a new official coordination structure will start in 2010. The INSPIRE-network is preparing a new SDI development strategy for years 2011 to 2015.

Box 3.1

The Finnish National Geographic Information Strategy (see also Finnish National Council for Geographic Information 2004 and www.paikkatietoikkuna.fi/web/EN).

The remaining countries have made different arrangements for the coordination of their national SDI activities. In four countries (Germany, Netherlands, Spain, and Switzerland), government interdepartmental bodies have been formally charged with creating national SDIs, which are now operational. In the Netherlands, the government took the lead and set up a new agency (Geonovum) to develop an operational national SDI. Belgium presents a special case, as two different agencies have come into being to coordinate SDI activities in Flemish-speaking Flanders and French-speaking Wallonia, and a national initiative began to emerge only recently.

Accession to the European Union—a unique historic opportunity to further the integration of the continent by peaceful means and extend the zone of stability and prosperity to new members (CEC 2000a)—has been a major driving force for SDI development in most of the central and eastern European countries. Ten countries—Cyprus, the Czech Republic, Estonia, Hungary, Latvia, Lithuania, Malta, Poland, Slovakia, and Slovenia—joined the EU in 2004, and Bulgaria and Romania joined in 2007. The EU has been assisting these countries in adopting EU laws,

Box 3.2

The Czech national SDI (see also Zimova 2009).

CZECH REPUBLIC

The development of the Czech national SDI within the country's broader national information infrastructure is an ongoing process. The main players are the Ministry of the Environment; the Ministry of the Interior; the Czech Office for Surveying, Mapping, and the Cadastre; the Czech Environmental Information Agency (CENIA); the Czech Association for Geographic Information (CAGI); and the GI cooperation forum "Nemoforum" that was set up in 1999 as a result of a EU-funded PHARE project led by the Dutch Cadastre.

Nemoforum produced a plan for the development of the Czech National Geoinformation Infrastructure from 2001 to 2005, which was accepted by the Czech government as background material for state information policy. Though never implemented, it has created political momentum for the new geoportals, especially those operated by the Ministry of the Environment, and the Czech Office for Surveying, Mapping, and the Cadastre. The majority of regional administrations as well as expert institutions also provide map viewing services. The new era began with transposition and implementation of the INSPIRE Directive, which created badly needed political support and momentum. A change of focus from mere map provision to georeporting also occurred, providing users with maps in legal and thematic context.

and it has also provided financial incentives to improve their infrastructures and economies. The process of EU enlargement can be seen as both a carrot and a stick for SDI development. The carrot is the need to develop an effective means of monitoring the spatial impacts of a wide range of social, economic, and environmental policies that are associated with EU accession, while the stick is the need to take steps to modernise public administration in these countries to make this possible.

The examples of Finland, the Czech Republic, Germany, France, and Lithuania give some indication of the diversity of experiences in various European countries (boxes 3.1 to 3.5). Finland is classified as a data-producer-led SDI that involves users. The Czech Republic became a member of the EU in May 2004 and now has a

Box 3.3

Geodateninfrastrucktur Deutschland (see also Lenk et al. 2008).

GERMANY

Germany is a federal republic with 16 states. Each of these states has its own surveying and mapping service, land and property registers, and statistical data collection services. To improve coordination of geographic information, the federal government established an Interministerial Committee for Geoinformation (Inter Ministerieller Auschuss fur Geoinformationswesen [IMAGI]) in 1998. IMAGI recognised that the development of the German national SDI (GDI-DE) required the cooperation of the 16 states and the local authorities within these states. Its efforts were given a substantial boost in February 2001 when the German Parliament passed a resolution on the "use of geoinformation in Germany." Following this resolution, IMAGI developed an implementation strategy, which included proposals to bring together existing concepts and strategies within Germany as well as harmonising metadata services and moving toward the creation of a national geodatabase. The Federal Agency for Cartography and Geodesy has played an important part in these developments. IMAGI is also working with the German national GI association (DDGI) and the private sector to promote the exploitation of geographic information. By the end of 2009, the federal government and most of the states had transposed the provisions of the INSPIRE Directive into federal or state laws.

formal mandate for SDI development. Germany and France are both classified as non-data-producer-led countries. Germany is a federal republic with a formal mandate for operational SDI development, whereas in France SDI development is only partially operational at the national level, with a number of well-developed examples at the subnational level. Lithuania is one of the former Baltic states of the Soviet Union that became an EU member in May 2004. It differs from the other European states in that it commissioned external consultants to carry out a feasibility study of the prospects for its SDI as part of an EU-funded initiative.

The reports prepared by SADL also contain a comparative evaluation of each country's experiences with respect to a number of key variables. These include organisational issues, legal issues

Box 3.4

SDIs in France (see also Salge et al. 2009).

FRANCE

The French national SDI is now operational, at least as a new Web service: www.geoportail.fr for the viewing service, and www.geocatalogue.fr for the metadata service. The government is considering the new shape of its national GI Council (Conseil National de l'Information Géographique [CNIG]) to promote the use of geographic information in France first established in 1985. The new CNIG will have about 30 members with a better balance between central government departments, representatives from local government, industry and the trade unions, and not-for-profit organisations. It will be chaired by a local government-elected person. The purview of the French national mapping agency (Institut Géographique National [IGN]) has been enlarged since 1999 as a result of the recommendations of the Lengagne report on "the evolution of geographic information in France and its consequence for IGN." IGN has the responsibility for the development of a large-scale basemap for the whole country (Referentiel Géographique a grand Echelle [RGE]), including orthophotography, topography, land parcels, and addresses. This has been completed for the whole country and the overseas department since 2007. In coherence with the national SDI, an ever-increasing number of local SDIs have come into being, including those in Lyon, Nantes, those in less-populated communes at the municipal level such as Haute Savoie and Vendée, those at the provincial level (départment), and those at the regional level such as Provence-Alpes-Cote d' Azur or Nord Pas de Calais. Thematic SDIs have also developed in various fields such as the wood industry, mountain economy, water, and biodiversity.

LITHUANIA

A detailed feasibility study of the prospects for an SDI in Lithuania was carried out in 2004 by external consultants within the context of the EU PHARE programme. Its main objective was to explore the issues involved in creating "an open shared infrastructure for accessing and distributing information, products, and services online." The findings of this study highlight the extent to which the Lithuanian Geographic Information Infrastructure (GII) is not just about technology but also requires a clear framework of agreements between government agencies and between the government, the private sector, and the citizens. These agreements will require political support for the GII at the highest level. They also show that substantial efforts will be required to raise current levels of awareness of the benefits of implementing an SDI within the stakeholder community in Lithuania. The consultants also produced detailed proposals for GII implementation from 2005 to 2008 and estimated that this will cost nearly €16 million. In light of these recommendations, SDI development is proceeding on three main lines: the establishment of the organisational framework for the GII, the creation of training materials to assist in capacity building with the Lithuanian GII community, and development of the necessary framework data for the GII. Work on the first of these goals started in 2005 supported by EU structural funds. Applications have been made for funding the other two activities.

Box 3.5

Lithuania's Geographic Information Infrastructure (see also Urbanas 2008).

and funding, fundamental reference data, and metadata services. The existence of some form of provision in most countries to cover the organisational aspects of SDI development and the advances made in the creation of fundamental datasets and metadata services contrasts sharply with the limited progress achieved in resolving the legal and funding issues associated with SDI development.

There is also a growing interest in the development of subnational SDIs throughout Europe. A workshop on "Advanced Regional SDIs" organised by the EU Joint Research Centre in May 2008 considered examples from Belgium, the Czech Republic, France, Germany, Italy, Spain, and the United Kingdom (Craglia and Campagna 2009) and a project cofunded by the EU eContentplus programme (www.esdinetplus.eu) has promoted cross-border dialogue about subnational SDIs within Europe. As part of this project, 12 workshops were organised in the main regions of Europe, and the experiences of nearly 160 SDIs have been explored. As a result of these activities, 13 SDIs from Denmark, France, Germany, Italy, the Netherlands, Norway, Portugal, Spain, and the United Kingdom were invited to take part in an SDI Best Practices awards ceremony in Turin in November 2009 (Masser and Rix 2009).

The following three case studies illustrate the main components of SDIs that were described in the previous chapter. The examples from the United Kingdom, the Netherlands, and Spain represent innovative and imaginative approaches to SDI development and implementation. The fact that none of the three SDIs is wholly national in character highlights the multilevel nature of SDI development and the issues that must be resolved at the subnational level during SDI implementation. Much of the data is collected at the subnational level, which is where choices often have the greatest direct impact on day-to-day SDI functions.

WORKING FROM A SHARED VISION: NORTHERN IRELAND'S GEOGRAPHIC INFORMATION STRATEGY

The first component of SDI development consists of the institutional arrangements that need to be made from the outset. These should take account of both the day-to-day coordination of SDI activities and matters of overall SDI governance. The extent to which the various groups of stakeholders are able to participate in the strategic decision making is particularly important. This presents problems in many countries because of the very large numbers of organisations that have a stake of some kind in SDI development and implementation. The challenge is how to ensure that these stakeholders develop a sense of shared ownership that would promote their continuing commitment.

One of the most distinctive features of the 2003 GI strategy that emerged in Northern Ireland was the innovative nature of the process that was used to build a shared vision of the strategy among key stakeholders from the outset. Ordnance Survey of Northern Ireland (OSNI) and its parent ministry, the Department of Culture, Arts, and Leisure, decided that a new approach to the development and implementation of geographic information policy in Northern Ireland was required. They chose the Future Search method, which helps participants in multistakeholder situations find common ground (box 3.6).

Box 3.6

The Future Search method was used in Northern Ireland to help stakeholders find common ground (see also Weissbord and Janoff 2000).

FUTURE SEARCH

Future Search is a unique planning meeting used worldwide by hundreds of communities and organisations to achieve two goals: (1) to help large, diverse groups discover common values, purposes, and projects; and (2) to enable people to create a desired future together and start working toward it right away. Typically, a Future Search involves a group of 60 to 70 people with many different perspectives. The size of the group allows for dialogue at every stage in the process. The optimal length for a Future Search meeting is two and half days with a minimum of four half-day sessions. The meeting facilitator has no direct involvement in the issues being discussed.

The Future Search method has five stages: (1) establishing the common history of the participants; (2) mapping the world trends affecting the whole group; (3) evaluating the progress made so far by each of the stakeholder groups; (4) considering some ideal future scenarios, indentifying the common ground that appears in each scenario, and confirming a common future; and (5) signing up to work together on action plans. The Future Search method avoids conflicts and focuses on the evolution of a shared agenda. Participants treat problems and conflicts as information rather than action items while searching for common ground and desirable futures. At the end of a successful Future Search, participants will have created desirable future scenarios and committed themselves and their organisations to action plans.

More than 50 organisations were invited to participate in a GI policy Future Search on the island of Lusty Beg in County Fermanagh, Northern Ireland, in February 2002. In addition to all the main stakeholders in Northern Ireland, British and European organisations sent representatives. Participants were divided into six more-or-less equal groups: (1) GI industry—technical; (2) GI industry—systems and data; (3) culture, arts, leisure, and tourism; (4) agriculture and environment; (5) emergency services, health, and transport; and (6) land property and networks.

The Future Search process worked well. The participants collectively created a mind map with 32 main trends and an even larger number of subtrends. They then identified and voted (privately) for their main priorities within this mind map. The already complete digital topographic data coverage of Northern Ireland emerged as the key factor, closely followed by the need to take account of the growing pressures on public funding. Other factors that rated highly included the recognition that GI is an economic resource, the need to promote environmentally sustainable development, and concerns about the lack of standardisation.

Key elements of a common ground for the future emerged during subsequent discussions: the importance of creating an overall GI strategy for Northern Ireland, the need to facilitate access and promote awareness, and the importance of partnerships in realising these objectives. On the basis of these discussions, all the participants agreed to work together in various working groups to hammer out the details.

The results of these debates were described in the consultation paper "Geographic Information Strategy for Northern Ireland," published by OSNI (OSNI 2003). Its vision was "to provide the strategic leadership required for a practical, coordinated, and inclusive approach to improving the collection, funding, dissemination, and use of geographic information, in order to maximise the social, economic, and educational potential of this crucial component of the national information infrastructure resource." The paper put forward a framework for SDI development and set up working groups for key sectors with immediate interests in better coordination: public safety/emergency services, land/property, transport, environment, utilities/networks, statistics, education/awareness, and key datasets (the activities of the last group cut across those of all the other groups). All these sectors were represented at Lusty Beg.

This resulted in the setting up of "a suitably robust and high-level strategic framework" to manage and coordinate the successful implementation of the GI strategy. The strategy was approved by the Minister of Culture, Arts, and Leisure with the support of the e-Government Board in 2004. A small GI support office with two-and-a-half staff positions was set up to provide a focal point for coordination. It was administered by a steering group consisting of the chairs of each of the working groups, the chair of the Northern Ireland branch of the Association for Geographic Information (AGINI), and the chief executive of OSNI. This group reported to the head of the Department of Culture, Arts, and Leisure in the Northern Ireland government and also to the e-Government Board. The steering group was chaired by the chief executive of the Ordnance Survey of Northern Ireland (box 3.7).

IAIN GREENWAY

Iain Greenway was chief executive of the Ordnance Survey of Northern Ireland from 2006 to 2008. As the chief survey officer of Northern Ireland and director of operations in Land and Property Services, he is now responsible for all land registration and rate collection activity in the agency. He also oversees the development of the Geographic Information Strategy for Northern Ireland.

Iain holds an MA in Engineering from the University of Cambridge, and an MSc in Land Survey from University College London. He joined the Ordnance Survey of Great Britain in 1986. A variety of technical survey posts followed, including consultancies supporting land reform in Eastern Europe. After completing an MBA at Cranfield University, he rejoined Ordnance Survey of Great Britain in 1995, undertaking roles in strategic planning and pricing and in sales and marketing as well as conducting management visits to Swaziland and Lesotho. He subsequently undertook a secondment to Her Majesty's Treasury, working in the Secretariat to the Public Services Productivity Panel (PSPP). From 2000 to 2006, Iain was general manager of operations and mapping for Ordnance Survey Ireland.

Iain is a Chartered Surveyor (MRICS), a Fellow of the Institution of Civil Engineering Surveyors (FInstCES), a Fellow of the Irish Institution of Surveyors (FIS), and a member of the Chartered Institute of Marketing (MCIM). He currently serves as vice president of the International Federation of Surveyors (FIG), and was the head of the RICS delegation to FIG from 1998 to 2006. He is chair of the FIG Standards Network and the FIG Task Force on Institutional and Organisational Development. He is also a member of the management and editorial boards of the journal *Survey Review.* He has published a range of articles and papers on geodetic surveys, business and management practices, sales and marketing, and standardisation.

Box 3.7

Iain Greenway, Chief Survey Officer of Northern Ireland and Director of Operations in Land and Property Services. Photo courtesy of Ordnance Survey Northern Ireland.

The working groups included representatives from the Northern Ireland government and universities, the utilities, and private-sector companies. There were also several representatives from UK agencies and from universities.

In 2006 the restored devolved Northern Ireland government undertook a wide ranging reform programme which included the creation of a new Land and Property Services (LPS) agency. This brought together under one roof the Ordnance Survey of Northern Ireland, Land Registers Northern Ireland, the Valuation and Lands Agency, and the Rate Collection Agency, on the assumption that an integrated set of land and property services would promote economic development. The agency became part of Northern Ireland's Department of Finance and Personnel (DFP).

Once the organisational changes were complete, LPS initiated a review of the 2003 strategy on the basis that many elements of it had been completed and that the INSPIRE Directive and UK Location Strategy had been finalised. A series of consultation processes with key stakeholders (in Northern Ireland and beyond) produced a new Geographic Information Strategy for the period from 2009 to 2019, which was published in 2009 after approval by the Minister for Finance and Personnel and endorsement by the Ministerial Executive (cabinet). Its vision is to "improve services and thereby develop the economy, the environment, and the society of Northern Ireland by placing information about location at everybody's fingertips and supporting the development of sufficient skills and knowledge to exploit this information" (DFP 2009, 11).

This strategy builds upon the previous GI Strategy governance structure and aligns it with the UK Location Strategy governance structure (figure 3.1), on which Northern Ireland and the other UK devolved administrations are represented. Under this structure, the steering group has been replaced by a GI Delivery Board. This board will take advice from a newly created Northern Ireland GI Council and report to the Department of Finance and Personnel, which has overall responsibility for GI policy (Steenson 2009). A Northern Ireland INSPIRE forum consisting of Annex I and II dataset holders has been established and reports to the GI Delivery Board, along with a Stakeholder Forum for the Pointer address dataset for Northern Ireland, and sectoral forums (www.gistrategyni.gov.uk).

A key component of the strategy is the development of GeoHub NI, which provides a platform for discovering and sharing spatial data (www.geohubni.gov.uk and figure 3.2). This enables every public servant to access appropriate geographic information to facilitate policy development and evaluation, administration, and service delivery at the desktop in a seamless way. The GI Strategy work in Northern Ireland is funded by the Northern Ireland Mapping Agreement (NIMA), which came into force in 2006. Under the terms of this corporate supply agreement, all Northern Ireland public servants can use geographic information produced by Land and Property Services free at the point of use. This removes funding as a major constraint to the use of spatial data.

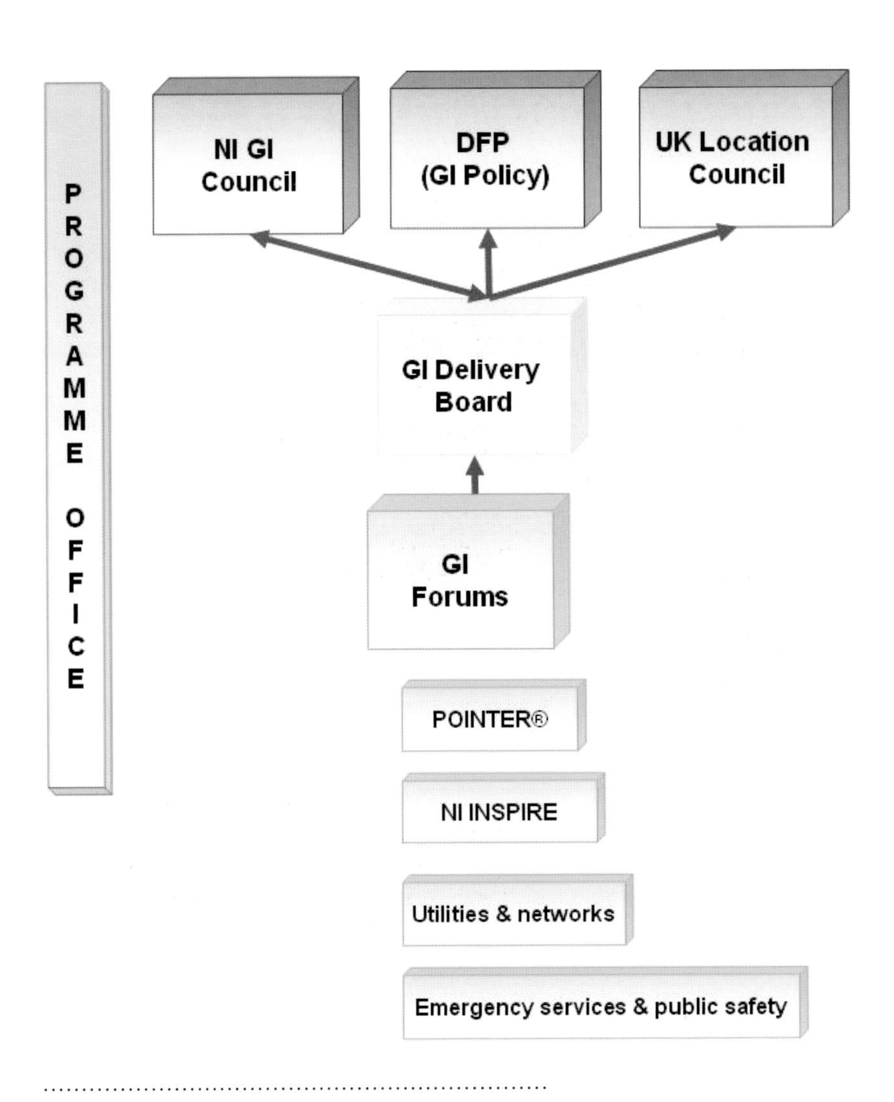

Figure 3.1

Organisational structure of Northern Ireland GI Strategy. Reproduced with permission of Land & Property Services Crown Copyright.

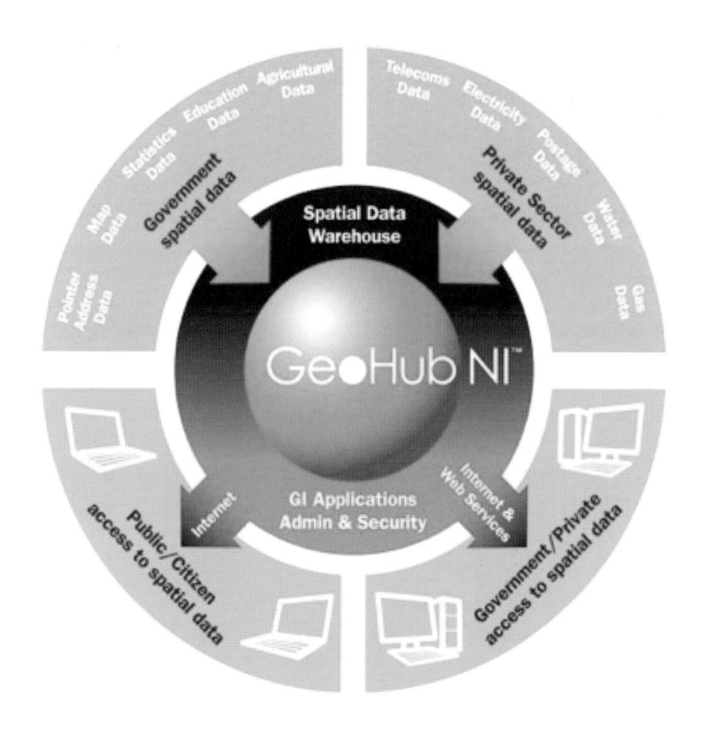

Figure 3.2

Northern Ireland's GeoHub. Reproduced with permission of Land & Property Services Crown Copyright.

FORMING COALITIONS TO CREATE DATA RESOURCES: THE DUTCH GBKN

The second component of SDI development deals with the creation of fundamental datasets. These datasets differ from the thematic datasets that are needed for particular kinds of applications such as environmental or transportation system modelling. Fundamental data typically includes geodetic reference points, elevation and hydrographic data, street names, addresses, administrative boundaries, and property and transportation information. Large-scale maps and property information must be seen as particularly important components of these fundamental datasets because they provide the basic building blocks for a lot of other fundamental data. In many European countries, the creation and maintenance of large-scale datasets of this kind lie outside the traditional purviews of national surveying and mapping agencies, and these challenges often call for innovative solutions.

The GBKN initiative (Grootschalige Basis Kaart Nederland [Large-scale Basemap of the Netherlands]) (www.gbkn.nl) is a good example of the kind of coalitions that must be built. The task involved the creation and maintenance of a large-scale map of the Netherlands at a scale of 1:500 to 1:1000 for urban areas and 1:2000 for rural areas. The coalition relies heavily on the continuing commitment of the participants, with the costs shared between central- and local-government agencies and private utility companies. The national joint venture agency (Landelijk Samenwerkingsverband [LSV]) that manages the project as a public–private partnership between data producers and data providers has been operating at the local, provincial, and national levels for more than 10 years.

The Dutch government approved the creation of GBKN in 1975 and assigned the practical work of database development to the Dutch Cadastre. The project progressed slowly due largely to the high financial input required from the Cadastre. To deal with these problems, a new partnership framework was launched in November 1992, with utility companies agreeing to pay 60 percent of the costs, and the Dutch Cadastre and the Association of Dutch Local Authorities (VNG) each agreeing to pay 20 percent (Murre 2002). Provincial public–private partnerships (PPPs) formed by the Dutch Cadastre, Dutch Telecom (KPN), and, wherever possible, regional utility companies and municipalities took on the responsibility for the implementation of the project.

The outcomes of these efforts were very positive. Nearly 60 percent of the whole country had been covered by 1995, and the last piece of the puzzle was fitted into place in January 2001. Some people called it "the map that had taken twenty five years to complete." But it probably would have never been completed at all without establishing a cost-sharing public–private partnership involving the key beneficiaries. Map production was also only the first stage of a longer process during which the map is regularly updated by the partners. Some of the discrepancies between the different pieces of the map have been resolved since 2001, and general guidelines have been developed to help the participants. Although some differences remain between the pieces, the GBKN is generally adequate for use as a framework layer in the Dutch national SDI. The

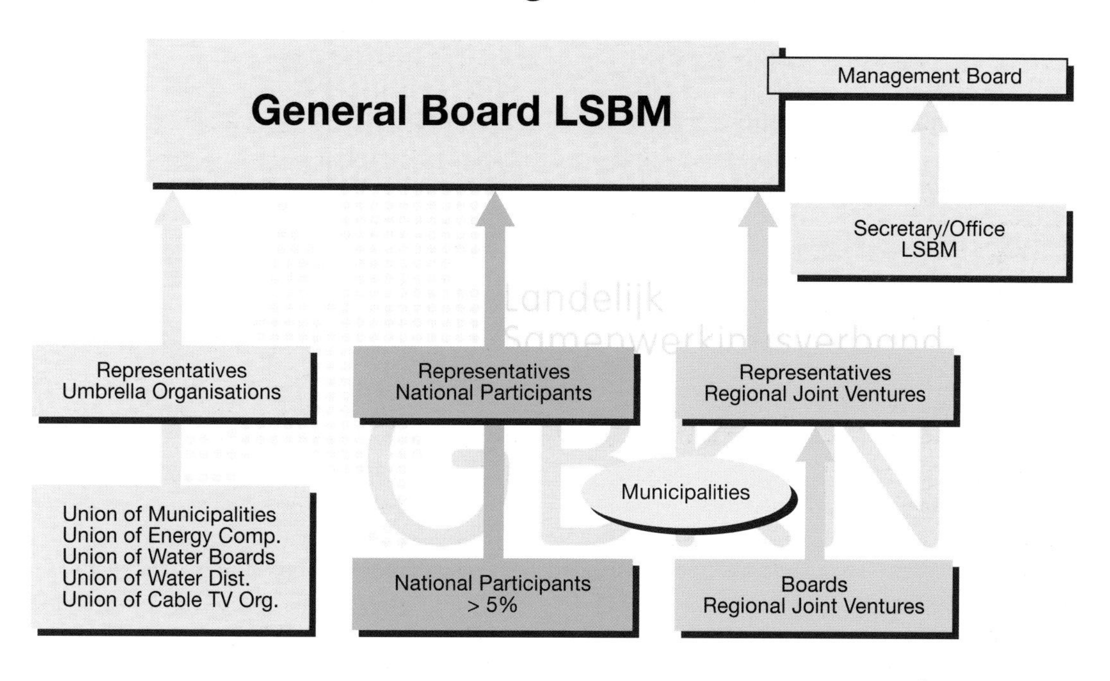

Figure 3.3

Organisational structure of GBKN. LSBM in the figure above means large-scale basemap. Courtesy of LSV GBKN.

map contains three types of information: hard topography (buildings, constructions, paved roads), soft topography (waterway boundaries such as hedges and fences), and identifiers (street names, house numbers, names of waterways).

A hierarchy of not-for-profit organisations with defined missions established under Dutch law *(stichtingen)* forms the framework of GBKN management. It is built around 10 regional organisations more or less corresponding to the Dutch provinces. The organisations have a shared mission "to produce and supply a digital large-scale map in order to support the business processes of the users as effectively as possible" (figure 3.3). The board membership of each of these bodies consists of representatives from regional utility companies, the Dutch Cadastre, and the municipalities. The national-level General Board includes representatives from regional organisations, user groups, the national stakeholders including the Dutch Cadastre, the Association of Water Boards, the national telecom company (KPN), and the Association of Dutch Municipalities. A small support centre based in the Dutch Cadastre's offices in Apeldoorn assists the General Board (box 3.8).

The initial production costs of the GBKN amounted to about €250 million, and maintenance costs were running at around €18 million a year in 2009. All the participants currently own the digital map, and the board operates a cost recovery policy for nonmembers of the joint venture. The LSV launched its own Web service in 2001 and received more than 20,000 hits per month during 2009 (figure 3.4).

Big changes are currently under way for GBKN as a result of the implementation of the Dutch government's e-Government programme to create basic registers for six key spatial datasets, including large-scale topographic data by 2014 (Peersmann et al. 2009). The implementation of these proposals requires the dismantling of the present public–private partnership structure and its replacement by a legally mandated public-sector body. The transition from the public–private partnership to the new body requires change in many domains, such as technical standards, production methods, IT architecture, organisational structures, legal framework, financial issues and user requirements. It is envisaged that it will be carried out in three stages, each of which will last for two years: (1) an initial policy design phase, (2) implementation and transition management for the producers, and (3) implementation and change management for the users.

Expected benefits from this restructuring include improved cooperation between government agencies by combining administrative and geospatial information, as well as the delivery of better services for citizens and the private sector. In the process, the GBKN will become the BGT (Basisregistratie Grootschalige Topografie), and the "spaghetti" structure of part of the current dataset will be upgraded into an object-oriented key registry.

Notwithstanding the changes that are currently taking place, some useful lessons can be learnt from past GBKN experience. Firstly, the original project would never have been completed without the willingness of very different organisations in both the public and the private sectors to share the costs of database creation and subsequent maintenance. Secondly, those organisations had to commit themselves to agreed standards. Thirdly, some form of top-level management was essential to ensure the smooth running of the project. The costs of this tier of management are minuscule by comparison with overall costs. Finally, the case study shows how the enlightened self-interest of the

MARTIN PEERSMANN

Martin Peersmann is general manager of the Large-scale Basemap of the Netherlands (GBKN). His primary responsibility is to execute the transition of this organisation from a public–private partnership to the legally founded Large-scale Topography of the Netherlands (BGT) key-registry as part of the Dutch GIDEON policy under the Ministry of Housing, Spatial Planning, and the Environment (VROM).

Martin Peersmann studied geology at the University of Utrecht before working as a production geologist for Shell, where he helped develop and execute its corporate geoinformation policy. In 1996, Martin moved to the Dutch organisation for applied scientific research (TNO) and became head of the Reservoir and Geoinformation Research group, where he worked on projects involving decisions support systems, smart fields technology, and CO_2 storage. In 2002, Martin became a member of the managing board of directors for the National Institute for Applied Geoscience of the Netherlands (NITG-TNO), where he managed the Dutch Subsurface Data and Information Programme (GIP) as part of the Dutch mining regulations on behalf of the Ministry of Economic Affairs.

Martin has been involved with the Dutch national SDI coordination agency (Geonovum), as well as EuroGeoSurveys and INSPIRE at the European level. In 2004, he became chair of the Users Advisory Council of the Space for Information (RGI) research and development programme financed by VROM. In 2007, he became chair of the National Data Repositories (NDR) Work Group and was involved in organising NDR9, the ninth in a series of global meetings promoting the collaboration of regulatory agencies and industry on oil and natural gas data management standards. NDR9 was hosted by the Directorate General of Hydrocarbons (DGH) of India.

In Martin's view,

> *The biggest challenge we face is to perform and manage a timed transition of the governance, production organisation, and the technical upgrade of the existing large-scale topography information to meet demands of the emerging information society during concurrent production operations.*

Box 3.8

Martin Peersmann, General Manager of the Large-scale Map of the Netherlands (GBKN). Photo courtesy of Martin Peersmann.

participants resulted in a win-win situation. Each participating organization obtained access to the detailed large-scale database it needs for its operations at a fraction of total cost!

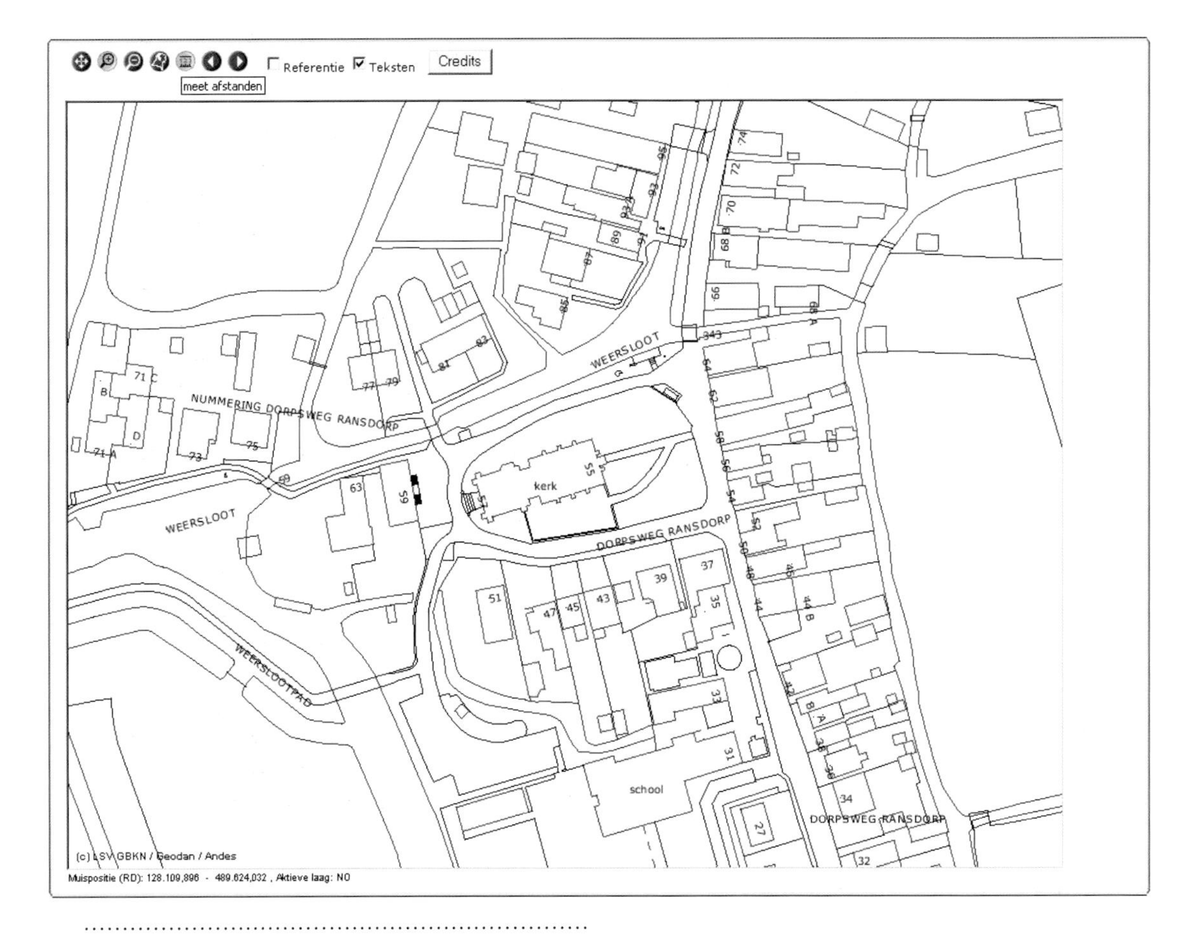

Figure 3.4

GBKN online. Courtesy of GBKN.

FACILITATING ACCESS TO DATA RESOURCES: CATALONIA'S IDEC

The third component of an SDI revolves around the development of metadata services and products. In order to maximise the use that is made of existing resources, potential users need to know what data exists for a particular area, as the same data may be potentially useful for many different users in the context of many different applications in a wide range of fields. They also need to know whether any limitations have been imposed on this data with respect to access. These can take the form of restrictions on reuse in commercial applications or charges for data provision. One feature of metadata products and services which has made them one of the success stories in the SDI field is that they can deliver results quickly to users with minimal resources.

The SDI for the autonomous region of Catalonia in Spain—Infraestructura Dades Especials de Catalunya (IDEC)—exemplifies the issues involved in the development and implementation of a wide range of metadata services (www.geoportal-idec.net). IDEC started as an initiative of the Catalan government's Secretariat for Telecommunications and the Information Society. Its objective is to promote the use of geographic information by making this data more available to public- and private-sector users and ordinary citizens. The project came under the direction of the Cartographic Institute of Catalonia and has been actively supported by the Catalan section of the Spanish national GI association (AESIG). Its main function is to develop an enabling platform to promote the dissemination of information and encourage contacts between data providers and data users. The project was also seen as a means of stimulating geographic-information-based projects in universities and research centres in the region.

IDEC began in 2002 with a two-year grant of €900,000 from the Catalan government and the European Regional Development Fund, which made possible the establishment of a small support centre with four technical staff to manage the IDEC. The Cartographic Institute of Catalonia provided the required technological infrastructure. The initial objectives were to compile information about existing data resources and products within the region and to create a software platform for making the data available to users throughout the region. IDEC first focused on data sharing within the departments of the regional autonomous government and then, in 2004, on data sharing within local governments in the region. The status of IDEC was formalised in December 2005 by a law passed by the Catalan Parliament, which established an independent support centre to manage IDEC within the Cartographic Institute of Catalonia (Guimet 2006). The support centre acts as the basic technical organisation for the promotion, exploitation, and maintenance of the Catalan SDI and the dissemination of associated data and services (box 3.9).

JORDI GUIMET

Jordi Guimet was born in Lleida, Catalonia, and studied industrial engineering in Barcelona. He was the deputy general director for Information Technologies in the General Directorate of the Cadastre Organization (Madrid, 1987–1991), where he was in charge of technical modernisation, introducing IT and GIS in all the cadastre offices and central services. He was also appointed regional cadastre director in Catalonia (Barcelona, 1992–2001). He was the director of IDEC from 2002 to 2005 and is now director of the Catalan Spatial Data Infrastructure support centre within the Cartographic Institute of Catalonia.

Jordi was awarded a doctorate in industrial engineering in 1990, and is an associate professor of information systems and business management at the Technological University of Catalonia. He is also director of the Master of Geospatial Technologies and Systems Programme at the Polytechnic of Catalonia. He founded the Spanish GIS Association in 1990 and was its president until 2000. He is the current president of the Catalan Association of GI Technologies (ACTIG).

In Jordi's words,

> *A great effort, especially in terms of organizational and institutional management skills, is required to set up a useful SDI. There is a necessary evolution process, from a data-centric approach to an applications approach, when the SDI becomes mature. After arriving at this stage, the number of possibilities, new applications, and new ways to cooperate created by the opportunities of sharing data, applications, and efforts increases very quickly, and all the investments and efforts make sense and their benefits are really worthwhile.*

Box 3.9

Jordi Guimet, Director of the support centre of the Catalan Spatial Data Infrastructure.

The support centre does not produce any data itself but rather facilitates data sharing between data producers and data users in the region. Its main tasks are to publish a catalogue of the data and services as well as to encourage data producers to make their data freely available over the Internet. It also provides a forum for representatives of organizations involved in geoinformation management to meet and exchange experiences. In this way it plays an important role in raising overall awareness of the potential of digital geographic information throughout the region.

The support centre has developed two main services to achieve its objectives: a searchable data catalogue service and a Web mapping service (WMS). The data catalogue service contains 28,000 records of datasets supplied by 133 different bodies, including government departments, municipalities, and private-sector organisations. The documentation of records conforms to the metadata standards approved by the International Standards Organisation (ISO) for dataset content, producer, spatial coverage, and currency and the conditions under which data can be obtained. This catalogue can be searched by topic, area, keyword, or coordinates (figure 3.5).

Since 2006 IDEC's Web mapping service has expanded dramatically. The number of data providers has increased from 12 to 130 (figure 3.6). As a result, users were able to access more than 5000 layers of reference and thematic

Figure 3.5

IDEC geoportal for Catalonia. Courtesy of IDEC.

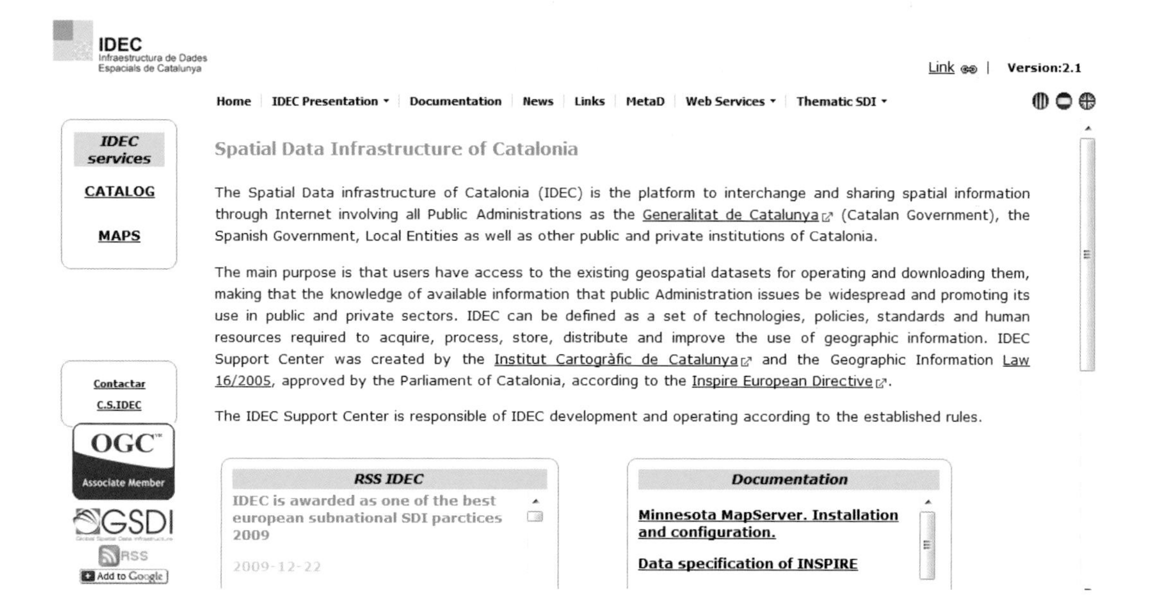

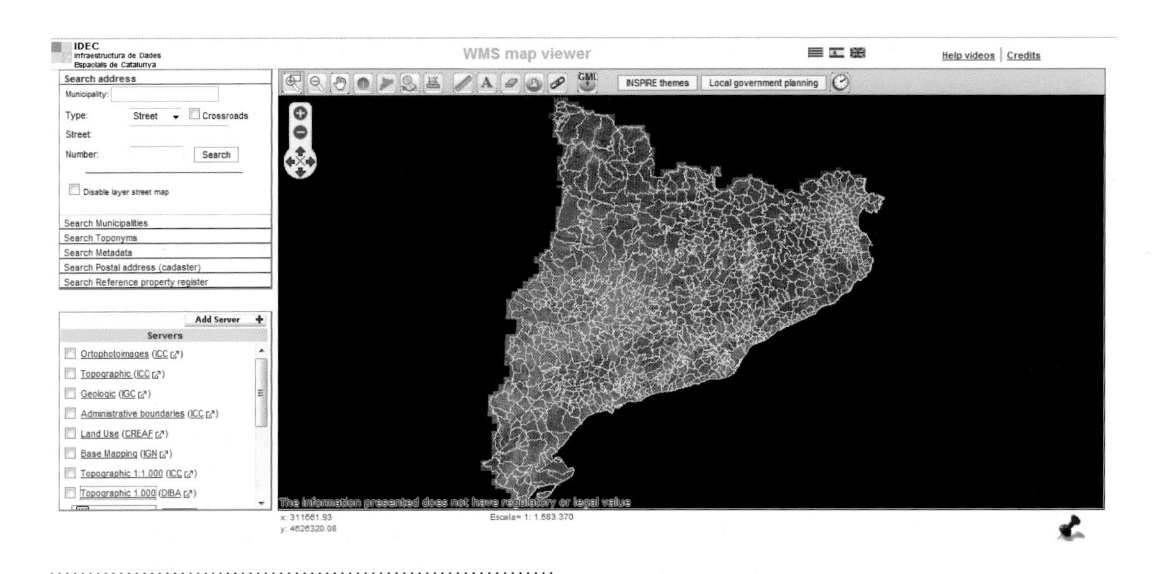

Figure 3.6

IDEC map viewer. Courtesy of IDEC.

data in 2009 as compared to only 150 in 2006. Available geoinformation includes topographic and environmental data, orthophotos, cadastral data, and urban planning information. Users can also download some of this data using Web feature service (WFS) and Web coverage service (WCS) software.

IDEC has made great efforts to increase the participation of municipalities. As a result, more than 400 municipalities made use of its facilities at the end of 2009. The funding of local GIS projects has added a strong capacity-building dimension to this work. Catalan e-Government funds have provided support to municipalities for creating metadata and publishing data in WMS and particularly for GIS projects meeting IDEC objectives. The centre offers support for applications based on its SDI resources platform. This includes customised services such as data catalogues, viewers, and feature editors.

A study of the socioeconomic impact of IDEC was carried out in 2007 by the Centre of Land Policy and Valuation of the Polytechnic University of Catalonia. This was based on a sample of 20 municipalities and 15 end-user organisations. The findings of this study suggest that the total investment costs of setting up IDEC between 2002 and 2005 would have been recovered in just over six months in 2006 by a variety of efficiency and effectiveness benefits to users (Garcia Almirall et al. 2008).

In the seven years that it has been in operation the number of data holdings accessible through IDEC's portal has increased substantially, and more and more municipalities are using IDEC to share their data. The support centre's role is primarily that of a facilitator in this process. In essence, its ultimate objective is to foster the sharing of data and services and maximise the use that is made of existing data resources in the broader context of an emerging information society. Initiatives such as IDEC can profoundly change existing administrative cultures by promoting greater cooperation between different levels of government as well as enabling the commercialisation of information gathered by public-sector organisations. The resources required for this kind of work are relatively trivial when compared to those needed for basic data collection and updating, and, as the impact study shows, the benefits greatly outweigh the costs of the initial investment.

CONCLUSIONS

The SDI phenomenon has spread throughout almost all of Europe, but countries vary considerably in the degree of progress made in establishing operational SDIs and in the types of approaches toward SDI development. The INSPIRE initiative offers much-needed standards and guidelines, and the governments of all EU member states are actively participating in INSPIRE-related activities.

The case studies from Northern Ireland, the Netherlands, and Catalonia exemplify innovative and imaginative approaches to SDI development and highlight the multilevel nature of SDI implementation in practice. Much of the data is collected at the subnational level, where decisions often have the greatest direct impact on day-to-day SDI functioning. The three case studies also illustrate the importance of involving stakeholders in all aspects of SDI development.

Support centres play a vital role in the development and implementation of SDIs. The Northern Irish support centre acts as the hub for SDI-related activities in Northern Ireland. The GBKN's national support centre has both strategic and management functions with respect to the other regional and national participants. And the IDEC support centre has been very proactive in mobilising regional data resources. The amount of money involved in backing these activities is surprisingly small: currently only two-and-a-half staff in Northern Ireland and four staff each for GBKN and IDEC. However, all three support centres are closely linked to key players in their regions and are able to draw upon the resources of the much larger organisations if necessary. This can be seen in the relationships between the Northern Irish support centre and the Ordnance Survey of Northern Ireland, between the GBKN group and the Dutch Cadastre, and between the IDEC support centre and the Cartographic Institute of Catalonia.

The developments described in the three case studies must also be viewed from a wider perspective. The Ordnance Survey of Northern Ireland is a member of the United Kingdom Location Council (UKLC). GBKN has links with the Geonovum, which plays an important role in the development of the Dutch national geographic information infrastructure. And IDEC is an important player in Spain's national IDE. E-Government agencies are actively involved in the management of all three strategies, and they are also actively involved in developments at the European level, particularly with respect to the INSPIRE initiative, which is described in the next chapter.

Developing a European SDI Strategy: The INSPIRE Initiative

EUROPE WITHOUT BORDERS

One of the guiding principles of the European Union is that there should be free movement of people, goods, and services between the member states. In other words, there should be no frontiers in Europe. With this in mind, seven countries signed a treaty in the small Luxembourg town of Schengen in June 1985 to end internal border checkpoints and controls. Despite the subsequent extension of the Schengen treaty to include most of the EU countries as well as Norway and Iceland, national borders still exist and create real barriers to the delivery of many essential services. Many local bus and train services within Europe stop at borders. Fire and ambulance crews often have difficulty crossing national boundaries to help victims. And small and medium-sized enterprises are often reluctant to develop their businesses across borders because of language and regulatory barriers.

Given that Europe is composed of many countries, it is not surprising that a large proportion of Europe's population lives in border regions. However, border regions are unevenly distributed among European countries. For example, relatively small portions of large countries such as Britain, France, and Germany are eligible for funding through the European Commission's INTERREG IIIA initiative, while most or all of the land area of some smaller countries such as the Baltic states, the Czech Republic, Hungary, Slovenia, and Slovakia falls into the border zone category (figure 4.1).

National borders not only inhibit integrated development of the European territory; they are also reflected in differences in the methods that national surveying and mapping agencies in different European countries use to collect geographic information. For example, some United European Levelling Net (UELN) heights differ significantly from those used by national surveying and mapping agencies (figure 4.2). For Belgium, this difference is greater than two metres. National mapping agencies may also differ in how they represent objects on maps, the amount of detail and the terminology they use to describe these objects, and the currency of the data (figure 4.3).

All these problems have hampered the implementation of the European Commission's major multisectoral initiatives such as the Water Framework Directive (CEC 2000b), whose objective is to establish a framework for the protection of inland surface waters (rivers and lakes), transitional waters (estuaries), coastal waters, and groundwater. The implementation of the directive will ensure that all aquatic ecosystems reach a satisfactory status with regard to their water needs, terrestrial ecosystems, and wetlands by 2015. An important feature of this directive is that it requires member states to define river basin districts and prepare river basin management plans for each of these districts using GIS technologies. The

Figure 4.1

EU INTERREG IIIA regions eligible for cross-border project funding in 2004–2006. Administrative boundaries copyright EuroGeographics; map copyright European Communities, 1995–2006. Used by permission.

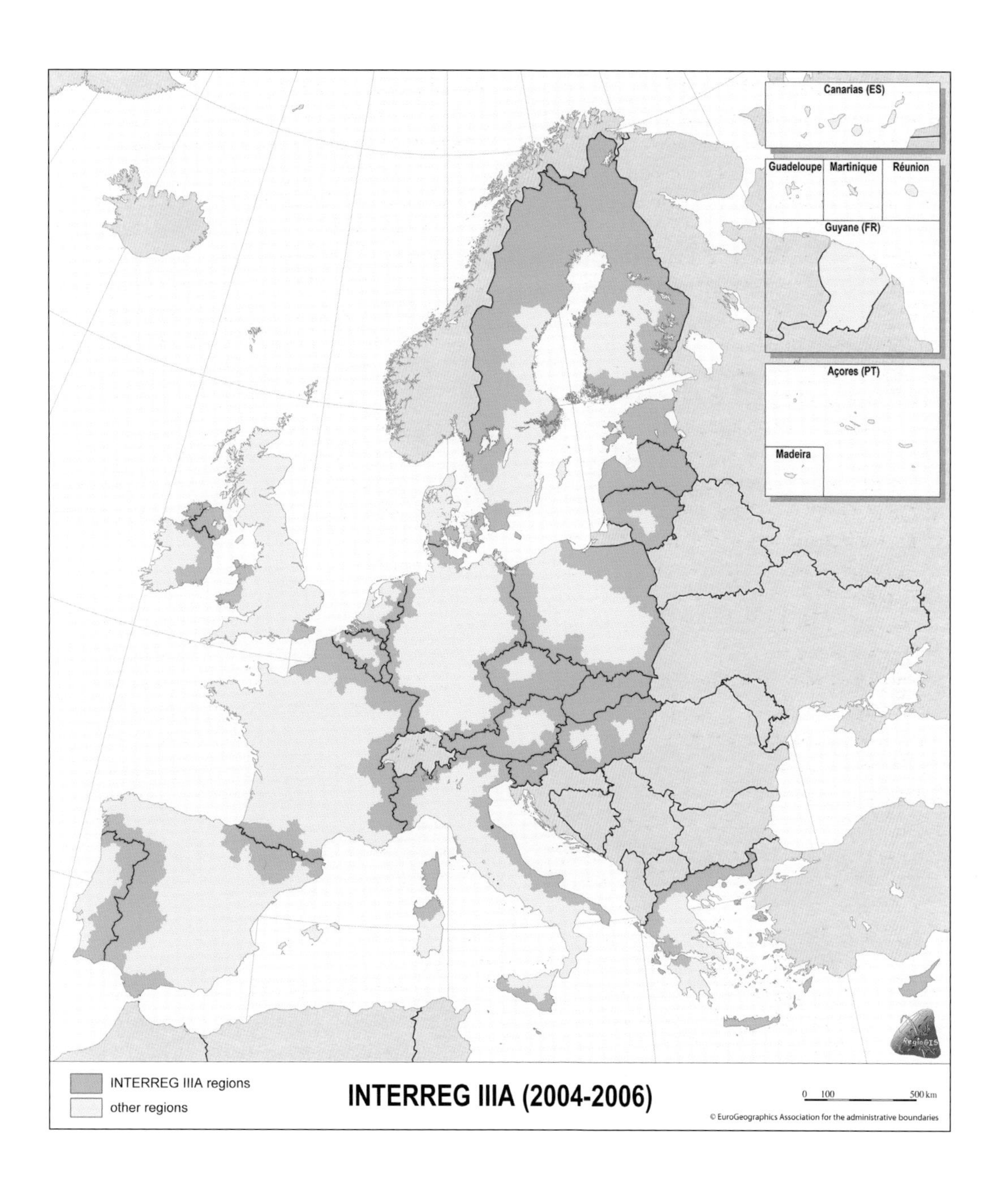
Canarias (ES)
Guadeloupe
Martinique
Réunion
Guyane (FR)
Açores (PT)
Madeira
RegioGIS
INTERREG IIIA regions
other regions
INTERREG IIIA (2004-2006)
0 100 500 km
© EuroGeographics Association for the administrative boundaries

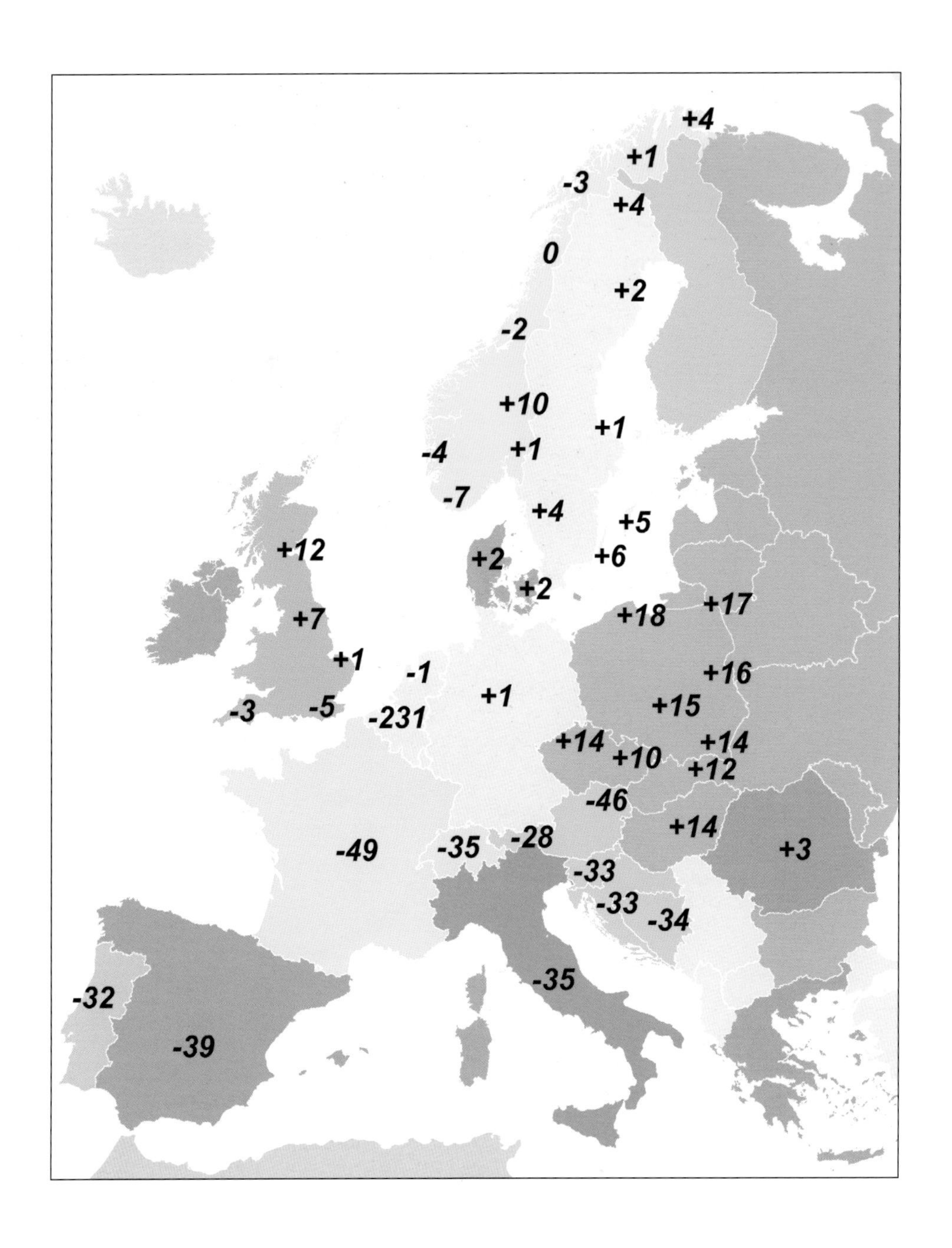
+4
+1
-3
+4
0
+2
-2
+10
+1
-4
+1
-7
+4
+5
+12
+2
+6
+2
+17
+7
+18
+1
-1
+16
+1
-3
-5
-231
+15
+14
+14
+10
+12
-46
+14
-49
-35
-28
+3
-33
-33
-34
-35
-32
-39

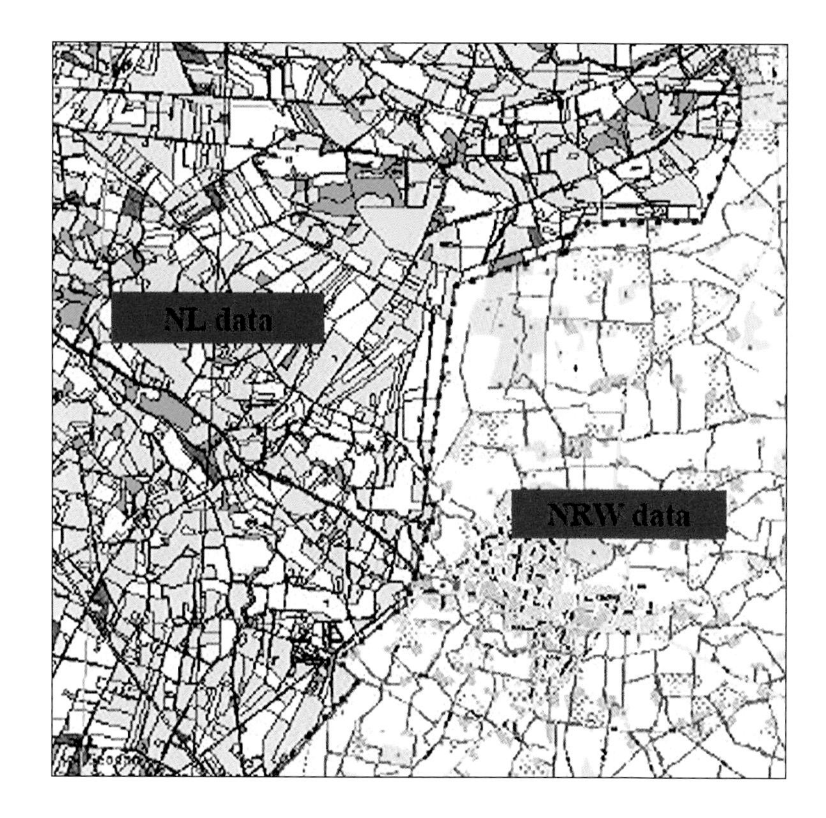

Figure 4.3

Composite map of a Dutch–German border region. Based on data courtesy of the Surveying and Mapping Agency of North Rhine–Westphalia (NRW) and the Netherlands (NL) Cadastre (Topografische Dienst Kadaster, Emmen).

Figure 4.2

Differences between United European Levelling Net (UELN) heights and national sea-level datum heights in Europe (cm). Courtesy of Bundesamt fur Kartographie und Geodasie.

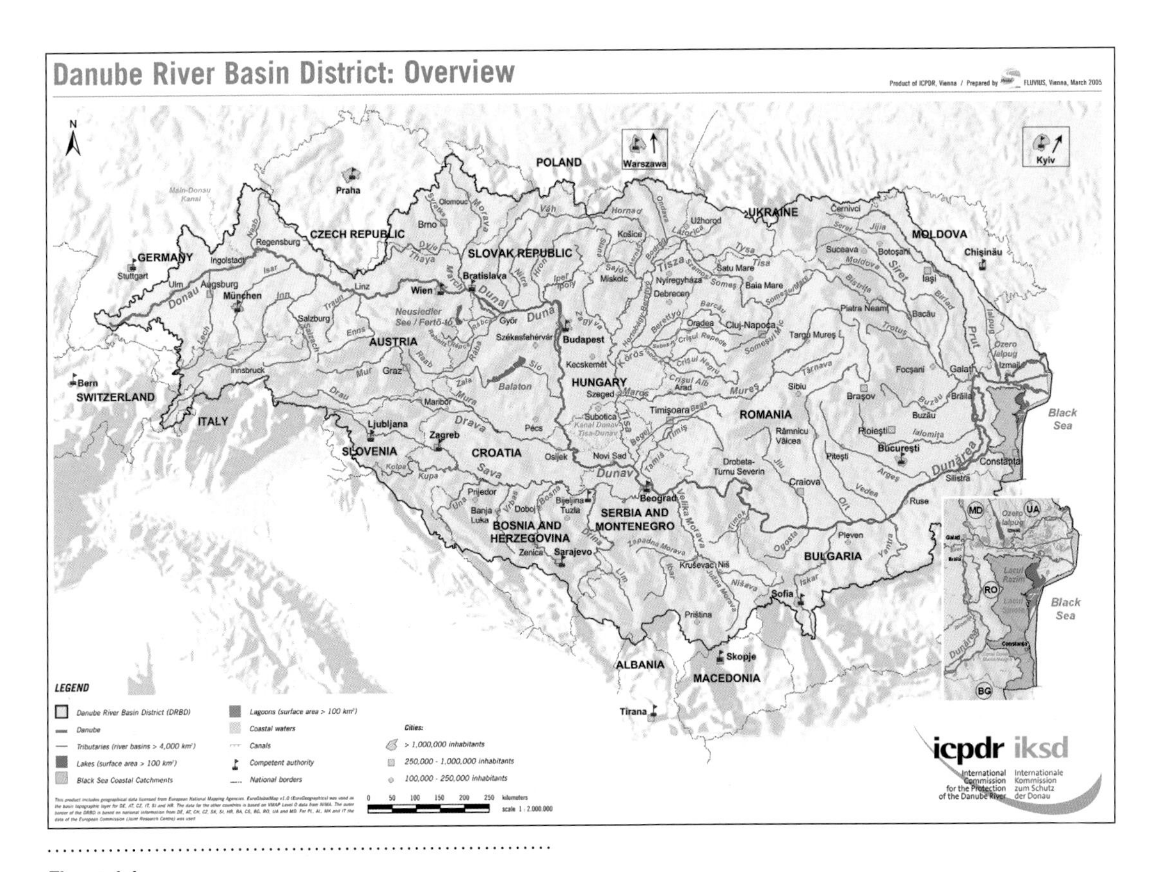

Figure 4.4

The Danube river basin covers 16 countries. Courtesy of International Commission for the Protection of the Danube River.

GIS applications will require the integration and harmonisation of large quantities of data derived from many different sources in the 27 member nations. In essence then, the directive requires not only the development of GIS applications but also a spatial data infrastructure to integrate the data into a unified framework.

The tasks involved in implementing the Water Framework Directive are formidable, as river basins do not necessarily correspond either to existing administrative divisions within Europe or to national boundaries. For example, the river Danube flows 2857 kilometres from its source in the Black Forest in Germany to the Black Sea (figure 4.4). Its basin spreads over 16 European countries, of which only nine currently are members of the European Union. Similarly, the river Rhine flows 1320 kilometres from its source near Reichenau in Switzerland to the North Sea. Its basin covers nine countries, of which seven are members of the EU.

Improvement of the quality of drinking water in Europe and conservation of the environment will require large-scale multisectoral management initiatives. They may also help to reduce the risk of floods, which have plagued many European countries over the centuries. The EU's directive on the assessment and management of flood risks (CEC 2007a) estimated that the 2002 floods of the Danube and Elbe rivers caused 700 deaths, displaced half a million people, and resulted in insured economic losses of at least €25 billion. It also pointed out that more than 10 million people live in areas of extreme risk along the Rhine, where potential flood damage is on the order of €165 billion. Flooding can also have severe environmental consequences, particularly when wastewater treatment plants or factories storing large quantities of toxic chemicals are involved.

The coastal regions of Europe face similar challenges. Nearly half the population of the European Union lives within 50 kilometres of the sea, and coastal-zone resources produce much of the region's wealth. However, increasing urban development and tourism along Europe's 89,000 kilometres of coastline have caused the degradation of many of Europe's coastal zones by reducing water quality, raising pollution levels, and accelerating erosion. Recent climatic changes have also increased environmental risks in these areas. These multifaceted problems require GIS-based strategies combining terrestrial and hydrographic data for integrated coastal-zone management (CEC 2001c).

INSPIRE

The development of the INSPIRE (Infrastructure for Spatial Information in Europe) initiative has been a complex process involving many groups of people since its inception in September 2001. This process can be divided into five stages: (1) antecedents, (2) initial development of INSPIRE to the end of 2002, (3) public consultations and impact assessments prior to publication of the draft directive in July 2004, (4) processes leading to the formal approval of the INSPIRE Directive in March 2007, and (5) INSPIRE implementation with particular reference to the development of implementing rules (IR) and the transposition of the directive into the legislation of the member states.

Antecedents (1985–2000). The first European initiative to make extensive use of digital geographic information was the Coordination of Information on the Environment (CORINE) programme, which was launched by the European Commission in July 1985 (box 4.1). It began as an experimental project for gathering, coordinating, and ensuring the consistency of information on the state of the environment and natural resources in the community, and later developed into a major resource for those concerned with environmental and related matters. One of its outcomes was the establishment of the European Environmental Agency in 1990 with a mission to provide "sound, independent information on the environment."

Environmental considerations have figured prominently in subsequent debates within Europe regarding access and dissemination of information. An EU Directive on Access to Information (90/313/EEC) was promulgated as early as 1990. This directive helped to encourage the collection and wider dissemination of environmental information and entitled the public to request and receive information on the state of the environment. A Convention on Access to Information, Public Participation in Decision-Making, and Access to Justice in Environmental Matters (The Aarhus Convention) was agreed on by the United Nations Economic Commission for Europe in 1998 to promote environmental democracy in the region (UNECE 1998). The European Community was a signatory to the convention and ratified it in February 2005.

The April 1994 publication of President Clinton's executive order establishing the national spatial data infrastructure for the United States (see chapter 2) prompted the Information Society Directorate of the European Commission to publish the first of a series of draft papers outlining its ideas for a European SDI in February 1995 (box 4.2). Economic and social matters rather than environmental considerations were the driving forces behind this initiative. These drafts, collectively known as GI 2000, were the subject of extensive consultations with the European geographic information community between 1995 and 1999 and helped to create favourable opinion about the development of a European SDI.

The public-sector information debates within the Information Society Directorate ran in parallel to those concerning GI 2000 during the late 1990s. The rationale behind these debates was the recognition that the public sector is the largest single producer of information in Europe and that the social and economic potential of this resource has yet to be tapped. Among the many types of information produced by the

COORDINATION OF INFORMATION ON THE ENVIRONMENT (CORINE)

The CORINE programme, adopted by the Council of Ministers in June 1985, was an experimental project for gathering, coordinating, and ensuring the consistency of information on the state of the environment and natural resources in the community. The European Commission's Directorate General for the Environment, together with national government ministries and national experts, was charged with coordinating data collection, processing and use issues, and building links to national land-cover programmes, many of which used both public agencies and private contractors, funded in part by the EU and directly by member state governments. In emphasising the development of geographic information systems for collating, standardising, and exchanging information about the environment CORINE drew attention to the complexity of dissemination and reuse of geographic information in digital form.

Box 4.1

The CORINE programme pioneered the use of geographic information systems for environmental applications in the European Union (CEC 1985).

GI 2000: TOWARD A EUROPEAN POLICY FRAMEWORK FOR GEOGRAPHIC INFORMATION

Discussions about a European geographic information infrastructure began at the meeting convened by the Information Society Directorate in Luxembourg in February 1995, which brought together key people representing GI interests in all EU nations. Meeting participants resolved the Directorate should initiate and support a widespread consultation process within the European GI community and prepare a policy document for the Council of Ministers. The impetus for this initiative came from a concern that "Europe is in danger of missing out on an important opportunity to exploit the potential of GI to contribute to social and economic development." National initiatives needed coordination to ensure that EU objectives could be met as well.

The final GI 2000 document, released in September 1999, recommended the creation of a high-level working party to provide leadership. The working party would include representatives from both the public and private GI sectors, including users, and it would be chaired and facilitated by the European Commission. Although never integrated directly into EU policy, the GI 2000 initiative did a great deal to create opinion within the European GI community that was favourable to the development of a GI strategy.

Box 4.2

The GI 2000 initiative helped to create favourable opinion toward an SDI for Europe.

public sector, geographic information stands out as having considerable potential for the development of digital products and services. The Reuse of Public Sector Information Directive was approved in November 2003 after extensive consultations with the key players in the field (box 4.3). The adoption of this directive is mandatory for the governments of all 27 EU member states, and they were legally obliged by the EU to incorporate its provisions into their respective national legislation by June 2005.

Eurostat, the statistical office of the European Communities, set up the GISCO (GIS of the Commission) project in 1992. From the outset GISCO encountered problems similar to those experienced by policy makers at the national level in obtaining the geographic information they need for decision making. In 1999 the European Commission set up an interagency Committee on Geographic Information (COGI) to increase awareness of the potential of geographic information within the commission and to coordinate its use and dissemination.

Initial development (2001–2002). An important driver behind the INSPIRE initiative was the Sixth Environment Action Programme of the EC for the period 2002 to 2012, which was approved by the European Parliament and the Council of Ministers in July 2002 (CEC 2002a). The programme identified four areas for priority action: (1) climatic change, nature, and biodiversity; (2) environment and health; (3) sustainable use of natural resources; and (4) management of waste. It called for policy making based on

Box 4.3

The EU Directive on the Reuse of Public Sector Information facilitated the sharing of geographic data with commercial entities (CEC 2003).

EU DIRECTIVE ON THE REUSE OF PUBLIC SECTOR INFORMATION

The 2003 EU Directive on the Reuse of Public Sector Information provided a framework for procedures dealing with requests for reuse of public-sector information, requests for reuse in all formats and languages, the transparency of conditions applicable to reuse, the need for a fee structure, and the availability of standard online licenses. The directive prohibited exclusive arrangements except where necessary for the provision of services in the public interest, and stipulated avoidance of cross subsidies between public and commercial parts of government agencies. The directive's measures are mandatory for all the EU member states, which had until July 2005 to incorporate the measures into their national legislation.

ALESSANDRO ANNONI

Alessandro Annoni graduated from the University of Milan, where he studied physics. He worked in the private sector from 1979 until 1996, managing companies specialising in advanced studies in geoinformation and earth observation and in GIS software development. He has been working at the European Commission's Joint Research Centre (JRC) in the Institute for Environment and Sustainability, in Ispra, Italy, since 1997. He is currently the head of the Spatial Data Infrastructures Unit, which includes a technical team of about 40 scientists working on spatial data infrastructures, the largest group in Europe in this field.

Alessandro has 30 years working experience in various environmental fields (e.g., forestry, agriculture, oceanology, hydrology, nature protection, and conservation) dealing with spatial planning, spatial analysis, environmental modeling, geoinformation and related technologies (such as GIS), remote sensing, image processing, system design, and software development. He has been the technical coordinator of the INSPIRE initiative since 2001. In 2006 Alessandro was appointed co-chair of the Architecture and Data Committee of the Group on Earth Observations (GEO) acting on behalf of the European Commission. Alessandro is a member of the Executive Committee of the International Society for Digital Earth (ISDE) and a member of the International Panel of the Centre for Earth Observation (CEODE) of the Republic of China.

In Alessandro's view,

> *INSPIRE is a directive establishing the legal framework for setting up and operating spatial information infrastructures in European Union member states.*
>
> *We have engaged hundreds of stakeholder organisations across Europe in the drafting of this legal framework. The complexity of this participatory approach is certainly innovative not only for SDIs but for European Union public policy in general. The resulting consensus-based policy and strong network of stakeholders is instrumental in the implementation of this distributed SDI. Creating a broad social network with empowered stakeholders and building on existing infrastructures, professional practices, and agreements are central features of INSPIRE. This approach entails multiple challenges, which we are striving to address together with our partners. INSPIRE is an interesting model for developing not only a technological infrastructure, but also shared practices and working methods, thorough collaboration, and partnership. It takes a lot of time and effort, but it is well worth it to achieve a shared sense of ownership of both process and outcomes.*
>
> *I personally believe that INSPIRE's architectural approach is suitable for multicountry situations in which spatial data infrastructures already exist in some form. For this reason, I have decided to play a more active role in GEO, offering lessons and practices from INSPIRE implementations for the benefit of a global society.*

participation, and reliable and up-to-date information to support and monitor environmental policies.

Following an initial expert meeting in September 2001, the commission was asked to prepare proposals for the establishment of an Environmental European Spatial Data Infrastructure (E-ESDI). Each EU nation was invited to nominate two experts—one with a background in environmental matters and another with experience in the geographic information field—to become the nucleus of the expert group that would develop and implement the E-ESDI initiative. This group, together with other experts from different sections of the European Commission and the international geographic information community, met in Vienna in December 2001 to discuss the proposals put forward by the commission (CEC 2001a) (box 4.4).

The proposals can be divided into three sets. The first sets out the objectives and scope of the E-ESDI initiative and "aims at making available relevant, harmonised, and quality geographic information for the purpose of formulation, implementation, monitoring, and evaluation of Community environmental policy-making and for the citizen" (CEC 2001a, 6). In this initial step toward a broader cross-sectional SDI that would provide information for decision making in diverse sectors such as agriculture and transport, the proposals set out for the first time what subsequently became known as the INSPIRE principles, which would govern the development of an SDI for Europe (box 4.5).

Box 4.4

Alessandro Annoni, Head of the Spatial Data Infrastructures Unit European Commission Joint Research Centre. Photo courtesy of Alessandro Annoni.

INSPIRE PRINCIPLES

- Data should be collected only once and kept where it can be maintained most effectively.
- It should be possible to combine seamless spatial information from different sources across Europe and share it with many users and applications.
- It should be possible for information collected at one level/scale to be shared with all levels/scales; detailed for thorough investigations, general for strategic purposes.
- Geographic information needed for good governance at all levels should be readily and transparently available.
- Easy to find what geographic information is available, how it can be used to meet a particular need, and under which conditions it can be acquired and used.

Box 4.5

The INSPIRE principles (http://inspire.jrc.ec.europa.eu/).
Courtesy of Commission of European Communities.

The first set of the proposals also aims at developing a community legal framework for an E-ESDI and establishing general requirements for coordination and monitoring as well as triggering discussion in the Council of Ministers and the European Parliament. Detailed technical specifications would be dealt with at a later stage.

The second set of the proposals specified the organisational structure for the initiative and assigned roles to the groups involved. Overall management is placed in the hands of the lead commission services working in close collaboration with a technical coordination group under the direction of the Joint Research Centre (JRC) in Ispra, Italy (CEC 2001a, 16).

The third set of the proposals identified five main activities for developing framework legislation:

- Common reference data and metadata, chaired by Eurostat
- Architecture and standards, chaired by the JRC
- Legal aspects and data policy, chaired by the Environment Agency for England and Wales
- Funding and implementation structures, chaired by Lantmateriet, Sweden
- Impact analyses, chaired by the Netherlands

The Vienna meeting participants agreed that working groups should be established to develop position papers on each of the above activities within a fixed period of time. They also agreed that a position paper on thematic user needs should be prepared by the European Environmental Agency. A separate baseline study of the SDI state of play in all the different European countries was also commissioned by Eurostat to ensure that the proposals for INSPIRE were based on a thorough understanding of existing circumstances throughout Europe. This was carried out by the Spatial Applications Division of the Katholieke Universitiet Leuven (SADL) in Belgium.

Meeting participants recommended more than 70 national experts for these working groups. To keep their activities within manageable proportions, a distinction was made between core and shadow working-group members.

After the Vienna meeting, the initiative was renamed INSPIRE. The three commissioners responsible for the Environment, Economic and Monetary Affairs (including Eurostat), and Research (including the Joint Research Centre) signed a memorandum of understanding (MOU) in April 2002, which set out in some detail the roles of each of these bodies in the first developmental phase of INSPIRE. This move broke new ground, as it was the first time that three commissioners had signed an MOU to jointly develop a legal framework (box 4.6)

Position papers were completed in October 2002, and the first report of the SDI state of play in all the European countries was completed in August 2003 (Van Orshoven et al. 2003).

Toward an INSPIRE Directive (2003–2004). The next stages leading up to the publication by the European Commission of its proposal for the INSPIRE Directive involved a number of consultations, an extended impact assessment, and a scoping study recommending revisions of the initial proposals.

A public Internet consultation took place between April and June 2003. Respondents were asked to give their views on 18 questions regarding the principles underlying INSPIRE and the main issues raised in the position papers prepared by the working groups. Some

HUGO DE GROOF

Hugo de Groof was born in Antwerp, Belgium, and studied geography and informatics in Ghent and Brussels. After graduating he worked as a software developer in the private sector before joining the Laboratory for Image Processing of the European Commission Joint Research Centre in 1986. Two years later, Hugo joined the Advanced Technologies unit, where he worked on synthetic aperture radar (SAR) data processing and land cover applications before moving on to the Monitoring Agriculture with Remote Sensing (MARS) unit where he became deputy head of the unit.

In 1998 Hugo joined the Directorate General for the Environment and was closely involved in the launch of INSPIRE in 2001. He has been an active participant in its development since that time. His duties include its follow up in DG ENV.

In his view,

> *Every piece of European environmental legislation is the fruit of many years of dedicated teamwork, in which national experts and commission officials develop a common vision and propose measures to render this vision concrete, before it is actually submitted to political scrutiny. The development and implementation of the INSPIRE Directive is a model example of these collaborative efforts. Without the enormous commitment of many working in the shadows, the directive would never have seen the light.*
>
> *INSPIRE is an unusual environmental directive which is not linked to one particular environmental medium or problem, and therefore it presents a challenge to all who have to promote and coordinate its implementation. However, few question its value once its goals and objectives are properly explained. Many of those involved in developing or governing spatial data infrastructures in their regions and organisations have embraced the INSPIRE Directive, and this growing community has become like a family, bound by strong ties through common objectives and a belief that by working together, fully respecting each other, we can do more and better for our society.*
>
> *In the years working on the coordination of INSPIRE measures, I have learnt that respect and modesty, even more than determination, are indeed the main recipes to successfully coordinating such a multinational and multidisciplinary initiative like INSPIRE, in which top-down and bottom-up have truly become one.*

BENEFIT	€ MILLION PER ANNUM
EIAs and SEAs	100–200
Environmental monitoring and assessment	100
Environmental protection	300
Environmental acquisition	50
EC projects	5–15
Trans European Networks	140
Spatial data collection	25–250
Risk prevention	120–400
Health and environment	350
Total	**1190–1800**

Table 4.1

Estimated benefits of INSPIRE. EIA, environmental impact analysis (INSPIRE 2003, 6). Copyright European Communities, 2007. Used by permission.

Box 4.6

Hugo de Groof, Chief Scientist, Research and Innovation Unit of DG ENV. Photo courtesy of Hugo de Groof.

185 individuals and organisations from a wide range of groups within both the public and the private sectors in different European countries participated in the survey. The responses indicated a very high level of support for the INSPIRE principles. A public hearing on INSPIRE was also held in Rome in July 2003.

An extended impact assessment of INSPIRE was also carried out during 2003. This analysis found that the overall costs of data harmonisation; the development of metadata services; and coordination—to EU and to national, regional, and local organisations over the first 10 years of INSPIRE—might be somewhere between €200 and 300 million per annum (INSPIRE 2003). These costs would be borne largely by the public sector, and they would be incurred mainly at the regional and local levels. Estimating the likely benefits of INSPIRE was more difficult, but even a partial assessment indicated that they would amount to somewhere between €1.2 and 1.8 billion a year, at least six times the estimated costs (table 4.1).

In light of these and other studies, two task forces were set up at the end of 2003. One of these had the mandate to reexamine the scope of INSPIRE and its measures, while the other sought to strengthen the extended-impact assessment. The work of these task forces led to reductions in the scope of INSPIRE and the number of priority datasets. As a result, the estimated costs were reduced to between €92 and 137 million and the estimated benefits to between €770 million and 1.15 billion.

The commission proposal for a Directive of the European Parliament and of the Council of Ministers establishing an Infrastructure for Spatial Information in Europe (INSPIRE) was finally published on 23 July 2004 (CEC 2004). The proposal for the directive contained an 8-page explanatory memorandum, 27 general provisions, 34 articles, and 3 annexes describing the various types of data involved.

The explanatory memorandum stated that the proposed directive,

> ...creates a legal framework for the establishment and operation of an Infrastructure for Spatial Information in Europe, for the purpose of formulating, implementing, monitoring and evaluating Community policies at all levels and providing public information... It focuses specifically on information needed to monitor and improve the state of the environment, including air, water, soil and the natural landscape... INSPIRE will not set off an extensive programme of new spatial data collection in the member states... It is designed to optimise the scope for exploiting the data that are available, by requiring documentation of existing data and implementation of services aimed at rendering spatial data more accessible and interoperable and by dealing with obstacles to the use of the spatial data.

The 27 general provisions clarified the directive's position with respect to previous community legislation and related projects such as the commission's joint initiative with the European Space Agency to develop services for Global Monitoring of Environment and Security (GMES) as well as the commission's own GALILEO satellite navigation project.

The 34 articles were divided into seven chapters: (1) general provisions, (2) metadata, (3) harmonisation of spatial datasets and services, (4) network services, (5) data sharing and reuse,

(6) coordination and complementary measures, and (7) final provisions.

Following its publication, the draft directive entered a decision process whereby it required approval by the European Commission (the EU's executive arm of government), the European Parliament (directly elected legislators from EU member states), and the Council of Ministers (heads of ministries from EU member states) before it could become law, and its measures had to be incorporated into the national legislation of each of the 27 EU member states.

Approval (2004–2007). Government legislation goes through a complex set of procedures that must be followed whereby the draft is reviewed and amended before passing into law. In the European Union, the European Commission submits draft directives like INSPIRE to the European Parliament and the Council of Ministers for approval.

European Parliament members are directly elected by citizens of the European Union to represent their interests. Elections are held every five years. The 732 members of the European Parliament do not sit in national blocks, but rather in seven Europe-wide political groups representing all views on European integration.

The council represents the interests of the member states and its meetings are attended by one minister from each of the 27 national governments. The choice of minister to attend each meeting depends on what subjects are on the agenda. For example, when the council is discussing environmental issues, the meeting will be attended by the environment ministers from each European country and it will be known as the "Environment Council."

Most EU lawmaking uses a codecision procedure whereby the European Parliament shares legislative power equally with the council.

The proposal for the INSPIRE Directive entered the codecision process soon after its publication in July 2004. The European Parliament adopted an amended version at its first reading in June 2005. The Council of Ministers approved a different version of the directive a few weeks later. The commission did not accept the council's version.

The main areas of disagreement between the council and the commission were the following:

- "The Commission does not agree that intellectual property rights held by public authorities should be among the list of grounds for limiting public access to spatial data.
- The Commission also does not agree that the possibility of limiting access should be extended to cover discovery services referred to in Article 18(1)(a) of the Commission proposal, since this would mean that the public would not even be able to learn of the existence of the data.
- The Commission maintains that the view services referred to in Article 18(1)(b) of the Commission proposal should be made available free of charge, and cannot accept the Council position allowing public authorities to charge and license for these services under certain conditions.
- The common position makes the obligation to avoid obstacles to data sharing, as well as the rules for ensuring harmonised conditions for Community institutions and bodies, subordinate to the right of public data providers to charge and license other authorities for their data. It is also vague about the obstacles to be avoided. It will therefore be ineffective in

achieving one of the key aims of the proposal, and could even have the effect of increasing obstacles to the sharing of data.

- Finally, while the Commission agrees that the provisions relating to data sharing do not affect the existence or ownership of public authorities' intellectual property rights, it does not see the need for this to be stated in the text of the directive. If such a provision is to be included, it should be made equally clear that these rights must be exercised in accordance with the other provisions" (CEC 2006, 3).

At the second reading of the directive in June 2006, the European Parliament made 36 amendments to the version approved by the council. Most of these amendments were accepted in full by the commission but not by the council.

When the council and parliament are unable to agree on a piece of legislation, it is put before a conciliation committee made up of equal numbers of council and parliament representatives. The conciliation committee finally reached agreement on INSPIRE on 21 November 2006.

The version of the directive agreed by the conciliation committee was formally approved by the European Parliament and the council on 14 March 2007 (CEC 2007b). Unlike the previous version, it contains no explanatory memorandum, and the number of general provisions has increased from 27 to 35. The main text is still divided into 7 chapters, but they contain only 24 articles instead of the original 32. Several changes have also been made in the annexes with respect to the priority given to the different types of data. With the exception of several new articles dealing with the main areas of disagreement between the council and the commission described above, the basic structure of the directive remains intact, and most of the changes consist of clarifications of items in the original text.

Implementation (ongoing). The implementation of the INSPIRE Directive is a complex process that will take at least a decade. The first stage, which took place through the end of 2009, was the development and approval of the implementing rules (IR) as well as the transposition of the directive's provisions by each of the 27 member states into their own national legislation. The following stages of implementation largely involve the creation of operational services and interoperable datasets as set out in these IR. According to the commission's own road map, this may not be completed until 2019.

The implementing rules procedure. The commission set up drafting teams to prepare IR for the directive. The chairs of these teams are Marcel Reuvers from Geonovum, Netherlands (metadata); Clemens Portele from GDI-DE, Germany (interoperability and data specifications); Jean-Jacques Serrano from BRGM, France (network services); Clare Hadley from Ordnance Survey, United Kingdom (data and service sharing); and Marie-Louise Zambon from IGN, France (coordination, and monitoring and reporting). Each team, composed of members of spatial data interest communities (SDIC) and legally mandated organisations (LMOs), addresses one of the five main elements noted above, which are set out in the INSPIRE Directive.

The three lead services of the European Commission for the implementation of the INSPIRE Directive are Directorate General Environment, Eurostat, and the Joint Research

DANIELE RIZZI

Daniele Rizzi has a degree in civil engineering from the Politecnico di Milano. He has spent most of his professional life working on geographic information, starting in the private sector and later in a regional public administration agency, CSI-Piemonte (the Piedmont Consortium for Information Systems) in Italy. He joined the European Commission in 1993, working for five years for DG Information Society, where he was responsible for research projects exploring how to deliver georeferenced information and multimedia content to mobile devices. Daniele also worked on projects supporting the preparation of the INSPIRE framework.

Daniele is currently head of Geographic Information System of the Commission (GISCO) in Eurostat, which he joined in 2004. The GISCO sector is responsible for the reference GIS of the European Commission, including basic pan-European topographic and thematic information layers. Its activities include map production and spatial analysis supporting Eurostat and other EC Directorates-General. GISCO is responsible, together with DG Environment and the Joint Research Centre (JRC), for the implementation of the INSPIRE Directive.

In Daniele's view,

> *There are times when the principle "the whole is greater than the sum of its parts" does apply, and this is definitively the case for spatial data infrastructures, where a coordinated effort really creates added value from what could be achieved through individual initiatives. From the European perspective, and as a citizen, I believe that INSPIRE will provide a substantial contribution for an improved knowledge of our world, supporting policy making but also providing everybody a better understanding of the environment we all live in. INSPIRE is an old dream on its way to becoming a reality, and I am proud to have been given the opportunity to provide my little contribution to this great and exciting challenge.*

Box 4.7

Daniele Rizzi, Head of the Geographic Information System of the Commission (GISCO) in Eurostat. Photo courtesy of Daniele Rizzi.

Centre (JRC). DG Environment is the overall legislative and policy coordinator for INSPIRE, while Eurostat is the overall implementation coordinator and supports the development of IRs on data and service sharing, and monitoring and reporting (box 4.7). The JRC acts as the overall technical coordinator of INSPIRE, and is responsible for the development of IRs for metadata, data specification, and network services as well as the development of the INSPIRE Geoportal.

The commission recognised that the development of implemention rules required the participation of a large number of stakeholders from the member states. To assist the drafting teams and to make the process as inclusive as possible, the commission built a network of spatial data interest communities (SDICs) throughout Europe. These SDICs bring together "the human expertise of users, producers, and transformers of spatial information, technical competence, financial resources, and policies, with an interest to better use these resources for spatial data management and the development and operation of spatial information services. Through their activities they drive the demand for spatial data and spatial information services" (http://inspire.jrc.ec.europa.eu/). SDICs may represent a common interest in data themes, spatial information services, or legal and procedural issues. They may represent different sectors of society or different geographical areas. They can also be seen as strategic partnerships within and between the public and private sectors. SDICs include both professional bodies such as the Association of European mapping, land registry, and cadastral agencies (EuroGeographics); the European Geological Surveys Association (EuroGeoSurveys); as well as national umbrella organisations such as the British Association for Geographic Information (AGI) and the Association Française pour l'Information Géographique (AFIGéO) in France.

SDICs operate alongside the legally mandated organisations (LMOs), which are formally charged with one or more elements of INSPIRE implementation. LMOs include both umbrella organisations such as GIS Flanders in Belgium, IMAGI in Germany, and organisations that cover limited components of INSPIRE, such as national mapping, statistical, and environmental agencies.

The commission announced an open call for SDICs and LMOs to register their interests and nominate experts for the drafting teams in the second half of 2004. By the end of 2009, 367 SDICs and 198 LMOs had registered their interests. These stakeholders are kept informed about developments, are given the opportunity to review and comment on INSPIRE deliverables, and are asked to test draft specifications. Regular progress reports on INSPIRE implementation are presented at the annual INSPIRE Conferences, and the materials under discussion are all published on the INSPIRE Web site.

The commission has devised a procedure for how SDICs and LMOs should participate in the work of the drafting teams, with SDICs contributing to the development and testing of the IRs, and LMOs reviewing these rules prior to public consultation (figure 4.5). To facilitate the implementation process, the commission set up a regulatory committee consisting of representatives of the member states in August 2007. The INSPIRE Committee offers opinions on draft IR proposed by the commission in the form of a vote.

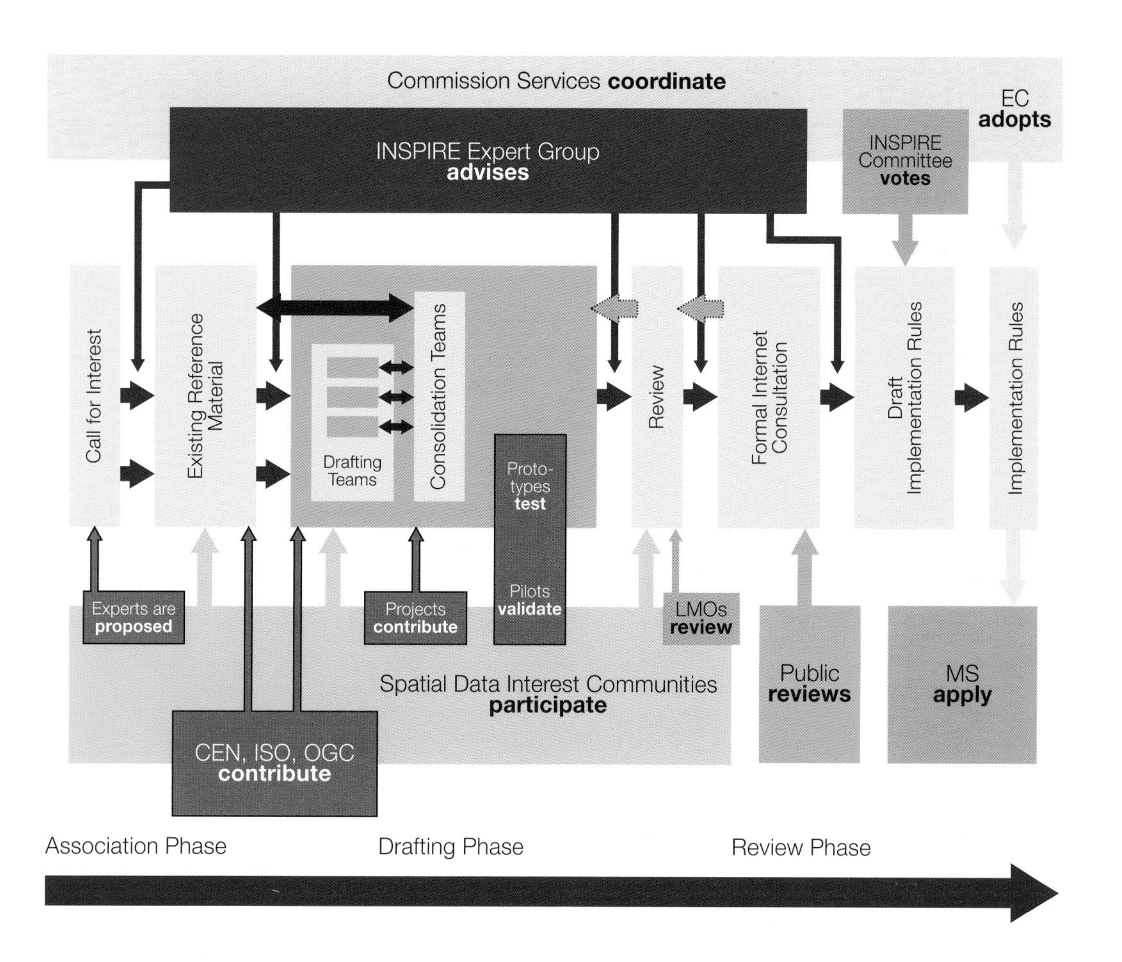

Figure 4.5

Interaction between the SDICs, the LMOs, and the drafting teams in preparing implementation rules for INSPIRE. Courtesy of Commission of the European Communities.

An example of the basic process for IR development can be seen from the work of the metadata drafting team, which released a draft of IR to the SDICs and LMOs in March 2007. More than 60 organisations responded, making more than 1000 separate comments. The drafting team revised the IR, and then sent another draft out for public consultation six months later. More revisions took place before the draft regulation for metadata was submitted to the INSPIRE Committee in May 2008. The committee unanimously supported the regulation, which was formally approved by the European Parliament and the council in December 2008 (CEC 2008a).

The Metadata Regulation sets out the definitions of the terms used, discusses the key elements, outlines instructions on multiplicity and considers the value domains involved. Key elements include the classification of datasets and services and the use of keywords, geographic location, temporal referencing, quality and validity, conformity, constraints related to access and use, as well as details of the responsible organisation and points of contact. An important feature of the regulation is that the directive requirements go beyond what is included in the core part of existing international standards while at the same time ensuring that they can be implemented in compliance with them. The drafting team also published more detailed technical guidelines based on the standards set out by the International Standards Organisation (ISO) in EN ISO 19115 and EN ISO19119 in October 2008.

Article 21 of the directive requires member states to monitor the implementation and use of their infrastructures for spatial information. Monitoring and reporting was the next set of IR to receive a positive opinion from the INSPIRE Committee in December 2008, following consultation with the SDICs and LMOs in March 2008. In June 2009 they were adopted as a Commission Decision addressed to member states (CEC 2009a). The text of the IR distinguishes between the tasks of monitoring and reporting. A quantitative approach is adopted for monitoring. This takes the form of eight sets of detailed indicators for each member state that describe their plans for the implementation of INSPIRE. Reporting is seen as a more general task that requires a qualitative approach. To meet the requirements of the INSPIRE Directive, member states are required to prepare reports every three years on five main topics: coordination and quality assurance procedures, the main stakeholders and their roles, the use of the infrastructure for spatial information, data sharing arrangements between public authorities, and estimates of the costs related to the INSPIRE Directive as well as examples of the observed benefits. Member states are required to submit their first reports to the commission by 15 May 2010.

Five categories of network services are defined in Article 11 of the INSPIRE Directive: discovery, view, download, transformation, and invoke (chaining) services. Consultation on the draft IR for discovery and view services began in December 2007. The revised IR were positively received by the INSPIRE Committee in December 2008 and formally approved by the European Parliament and council in October 2009 (CEC 2009b). The regulation sets out the search criteria and operations required for discovery services and the operations involved in the development of view services. It also requires member states to provide services that conform to the regulation by November 2011. To facilitate their introduction, the commission published detailed technical guidelines and a SOAP (simple object access protocol) primer with INSPIRE examples (Villa et al. 2008).

Consultations on the draft IR for download and transformation services began in February 2009, and the revised regulation received a positive opinion from the INSPIRE Committee in December 2009. This took the form of an amendment to the regulation approved in October 2009 with respect to download and transformation services, which consists of a short text and three annexes dealing with quality of service criteria, download operations and search criteria, and transformation operations, respectively.

The findings of a survey about the development of invoke spatial data service specifications were published in December 2009 (Lucchi and Millot 2009). According to the commission's current road map, IR for invoke services will be submitted to the INSPIRE Committee in June 2012.

An Initial Operating Capability Task Force was set up in June 2009 to help member states implement INSPIRE services and ensure interoperability with the INSPIRE Geoportal (figure 4.6). Initially, it will concentrate on the implementation of the discovery and view services.

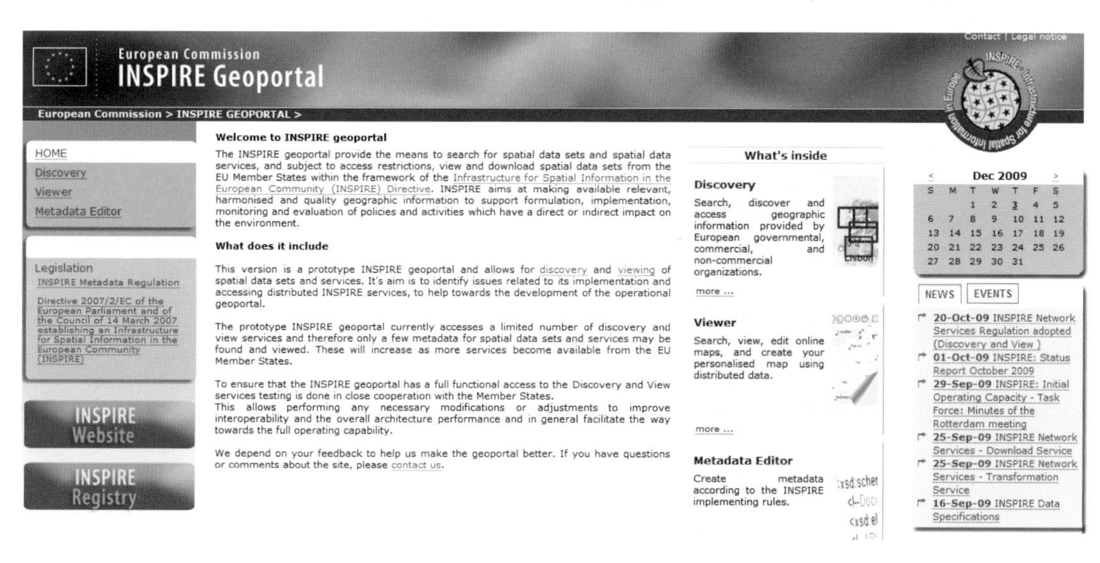

Figure 4.6

Homepage of the INSPIRE Geoportal. Courtesy of Commission of the European Communities.

Annex I
- Coordinate reference systems
- Geographical grid systems
- Geographical names
- Administrative units
- Addresses
- Cadastral parcels
- Transport networks
- Hydrography
- Protected sites

Annex II
- Elevation
- Land cover
- Orthoimagery
- Geology

Annex III
- Statistical units
- Buildings
- Soil
- Land use
- Human health and safety
- Utilities and government service
- Environmental monitoring facilities
- Production and industrial facilities
- Agriculture and aquaculture facilities
- Population distribution - demography
- Area management/restriction/regulation zones and reporting units
- Natural risk zones
- Atmospheric conditions
- Meteorological geographical features
- Oceanographic geographical features
- Sea regions
- Bio-geographical regions
- Habitats and biotopes
- Species distribution
- Energy resources
- Mineral resources

Table 4.2

INSPIRE data themes. Courtesy of the Commission of the European Communities.

Article 17(8) of the INSPIRE Directive requires the creation of IR to regulate the provision of access to spatial datasets and services from member states to the institutions and bodies of the community. The draft IR for data and service sharing were released for consultation in December 2008 and were subsequently revised before their submission to the INSPIRE Committee in June 2009, where they were positively received. The commission will consider their adoption during 2010. The regulation includes provisions regarding restrictions on access, arrangements concerning access, uses of spatial datasets and services, and response times (20 days) for member states to comply with written requests. Guidelines will most likely be published containing nonbinding agreements such as model contracts as well as illustrating related concepts such as framework agreements.

Annexes I, II, and III of the INSPIRE Directive give details of the 34 spatial data themes required (table 4.2). The most important difference between these categories concerns the timetable set out in the directive. The data specifications for the 9 Annex I spatial data themes must be completed by the end of 2009, while those for the 25 Annexes II and III themes, which mainly involve environmental data, are given until 2012 to resolve. Prior to the development of data specifications for the 9 Annex I reference data themes, generic conceptual models and encoding guidelines were established to provide a baseline for the work of the 9 thematic working groups set up for each of the themes. The draft guidelines for these themes were released for consultation in December 2008 and published in revised form in September and October 2009.

The draft regulations regarding interoperability of spatial datasets and services were positively received by the INSPIRE Committee in December 2009. These set out the technical requirements for the interoperability and, where practicable, harmonisation of spatial datasets and data services relating to the themes listed in the three annexes to the directive. The requirements took into account the findings of a user requirements survey as well as the testing results reported by stakeholders and materials from the member states regarding the costs and benefits of spatial data infrastructures at the regional level.

These regulations are more substantial than any of the previous regulations relating to the implementation of the INSPIRE Directive. They include a seven-page overview of the issues involved, followed by an annex dealing with the definition of common data types. The bulk the text contains details of the requirements for the 9 spatial data themes listed in Annex I of the directive (table 4.2). The sections dealing with coordinate reference systems, grid reference systems, geographical names, and protected areas are relatively brief (between one and three pages), while those dealing with administrative units, addresses, and cadastral parcels are slightly longer. Forty pages are needed to deal with the complexities of different kinds of transport network (e.g., air, cable, railway, road, and water), and another 20 cover issues relating to hydrographical data.

An INSPIRE registry has also been created to assist in the development of IR by the drafting teams, the thematic working groups for data specifications, and other participants in the

TRANSPOSITION OF THE INSPIRE DIRECTIVE

Various approaches have been adopted by EU member states to deal with the transposition of the INSPIRE Directive. Most countries have enacted legislation that directly transposes its provisions into statute law. A good example of this procedure is the Dutch law of 2 July 2009. This contains 6 chapters and 17 articles dealing with general provisions (articles 1-4), the obligations of administrative bodies (5-6), the rights of third parties (7), rules governing access to collections of spatial data and services related to spatial data (8-13), coordination and implementation (14-15), and closing remarks (16-17). A substantial explanatory memorandum accompanying the law contains an article-by-article summary of the links between the articles in the directive and those in the law.

An alternative approach has been adopted by the UK government based on the concept of statutory instruments. Unlike formal legislation of the Dutch kind, statutory instruments in the UK have the advantage of being direct legislation that does not require the formal approval of a national parliament. Statutory Instrument 3157 of 2009 contains the INSPIRE regulations for England, Wales, and Northern Ireland. It consists of 14 articles which deal with questions of interpretation and scope (1-4), intellectual property rights (5), metadata (6), network services and links (6-7), public access and charging (9-10), enforcement and appeals (11), data sharing between public authorities (12), internal complaints procedures (13), and coordination and monitoring (14). A separate guide to the INSPIRE regulations describes the obligations these regulations create for affected data holders, and an additional note compares the provisions in the regulations with each article of the INSPIRE Directive. Scottish SI 440 of 2009 covers the same topics for Scotland but includes a separate article dealing with the meaning of third-party rights. Both sets of regulations were published in early December 2009 and came into effect on 31 December 2009.

Box 4.8

Transposition of the INSPIRE Directive.

Pages of technical guidelines	
Metadata	74
Annex I	
Data specification themes	
Protected sites	115
Transport networks	202
Cadastral parcels	124
Geographical names	76
Geog grid systems	16
Coordinate reference systems	20
Administrative units	58
Addresses	177
Hydrography	175
Network services	
Discovery	17
View	36
Download	22
Transformation	17
Data and service sharing	*
Monitoring and reporting	59
Total pages	**1188**

Comments from consultations	
Metadata	1253
Data specification	
Methodology	1146
Encoding of spatial datasets	217
Annex themes and scope	1326
Generic conceptual model	1176
Spatial datasets and services	3649
Network services	
Discovery and view	787
Download services	392
Transformation services	149
Data and service sharing	508
Monitoring and reporting	359
Total comments	**10962**

** Not yet available*

Table 4.3

Some INSPIRE statistics.

consultation process. It contains the glossary and feature concept dictionary registers which form part of the ongoing process of IR development.

Transposition of the INSPIRE Directive. The INSPIRE Directive came into operation on 15 May 2007, and member states were given two years from this date to complete the tasks of transposing its provision into national legislation. Given the complexity of the legal procedures that are involved in such a process, it is not surprising to find that only Denmark and Portugal fully met this deadline together with some regions of Belgium and Germany that have devolved powers. By the end of 2009 nearly half the EU members had successfully transposed the provisions of the directive (eur-lex.europa.eu/). Member states differ with respect to the mechanisms used for transposition (box 4.8).

CONCLUSIONS

Transnational projects dealing with river basin management, the planning of coastal zones, environmental pollution, and other multisectoral issues require greater harmonisation of information of which geographic information is an important aspect. As Europe has many border regions, these considerations together with the requirements of the Sixth EC Environment Action Programme underpin the INSPIRE initiative, which calls for an unprecedented amount of cooperation between geographic data collection agencies in different countries and promises benefits up to ten times greater than the costs. To provide further support for such activities, the commission has launched an initiative to create a Shared Environmental Information System (SEIS) (CEC 2008b).

The task of creating the IR required for INSPIRE is a formidable one involving large numbers of participants from SDICs and LMOs in the member states as well as commission staff. Table 4.3 shows the scale of these operations with respect to the number of pages of technical guidelines and the number of comments received during the consultation processes. These highlight the role that the data specifications play in the implementation process as a whole and demonstrate the degree of stakeholder involvement in consultations that have played a key role in the open development of IR. This has led to the creation of a European-wide community of professionals with shared values and a common commitment to SDI implementation.

As a result of these efforts, most of the pieces of the puzzle represented by the first stage of INSPIRE implementation were in place or close to being in place by the end of 2009 with the exception of the data specifications for the 25 Annex II and Annex III data specifications, which should be submitted to the INSPIRE Committee by May 2012. The subsequent stages of INSPIRE implementation are largely in the hands of the member states. The current INSPIRE road map (November 2009) gives the following key dates for the member states: (1) metadata for the spatial data contained in Annexes I and II should be available December 2010, (2) operational discovery and view network services should be available November 2011, and (3) operational download and transformation services, and the newly collected and extensively restructured Annex I spatial datasets should be available on June 2012. December 2013 is the present target date for metadata for spatial data corresponding to Annex III, and January 2015 for newly collected and extensively restructured Annex II and Annex III spatial datasets. June 2015 and May 2019 are the present target dates for the availability of other Annex I and Annex II datasets, respectively.

Prospects for the Future

Recent years have brought about a great deal of progress toward SDI development and implementation, but a lot still remains to be done, particularly at subnational levels. The three case studies in chapter 3 are the exceptions rather than the rule in Europe. A more integrated approach to GIS applications and SDI development can bring considerable benefits to Europe's 100,000 municipalities, but many local officials have limited awareness of this potential and lack the basic skills needed to take advantage of SDI initiatives. The adoption of INSPIRE by the EU marks the beginning of a new phase of SDI implementation in Europe. INSPIRE is primarily an environmental initiative, and work in other key fields such as transportation, spatial planning, and agriculture is still at the early stages of development. Nevertheless, INSPIRE can be used as a model for multinational SDI development elsewhere in the world.

EMERGING TRENDS

The debate about spatial data infrastructures has reached a level of maturity that encourages reflection and facilitates critical evaluation of past experience. A big shift in thinking has taken place since the first generation of SDIs were launched throughout the world a decade ago. The "product model" that characterised most of the early SDIs has been giving way to the "process model" of the second generation (Rajabifard et al. 2003), reflecting a growing understanding of the concerns of data users as well as those of data producers. First-generation SDIs focused primarily on database creation, and as a result, most were led by data producers. The main driving forces behind the process model are (1) the desire to reuse data collected by a wide range of agencies for a great diversity of purposes, and (2) a shift from centralised structures to the distributed networks of the World Wide Web. The INSPIRE principles fully embody these changes in thinking (box 4.5).

A patchwork quilt or a collage. Many national SDI initiatives still seem to abide by the principle "one size fits all" and envision a relatively uniform product. However, SDI development has both top-down and bottom-up dimensions. National SDI strategies drive regional ones, and regional SDI strategies drive local ones. As most detailed database maintenance and updating tasks are carried out at the local level, input from local government has considerable impact on SDI implementation at the regional and national levels. The level of commitment to SDI implementation will vary considerably among regions and among municipalities. A national or multinational SDI will therefore likely be a collage, or a patchwork quilt, of similar but often quite distinctive components that reflect the commitments and aspirations of the different subnational governmental agencies (figure 5.1). While the top-down vision emphasises the need for standardisation and uniformity, the bottom-up vision stresses the importance of diversity and heterogeneity. Balancing the two visions will be a challenge, particularly for multinational initiatives such as INSPIRE, which also need to address the cultural aspects of SDIs that are already in place in different European countries.

Many countries are moving toward more inclusive models of SDI governance to meet the requirements of multilevel, multistakeholder SDIs (Rajabifard et al. 2006). The Northern Ireland case study in chapter 3 illustrates the

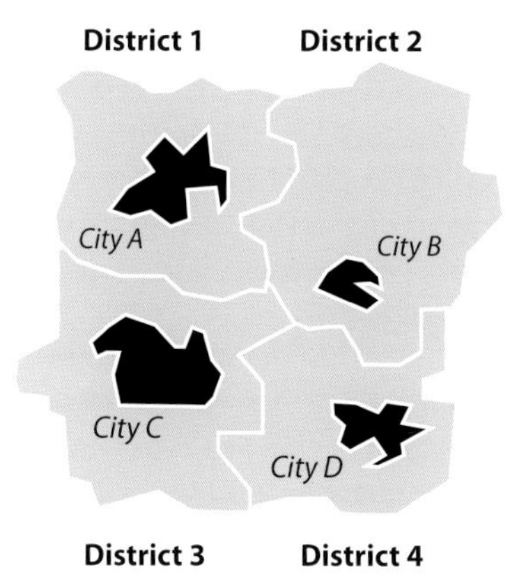

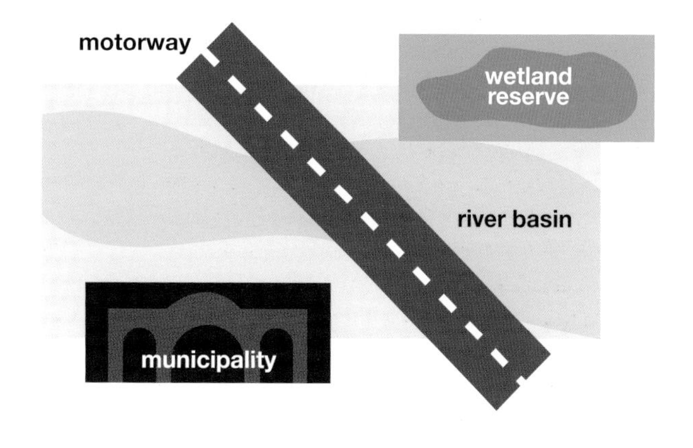

Figure 5.1

National and multinational SDIs may structurally resemble a patchwork quilt (left) or a collage (right).

institutional arrangements that must be made from the outset both for day-to-day coordination of SDI activities and for matters of overall SDI governance. Stakeholders must be able to participate in the strategic decision making. When very large numbers of organizations have a stake, creating a sense of shared ownership may be challenging, but their continuing commitment depends on it.

As a transnational SDI initiative, INSPIRE breaks new ground in striving to involve a large number of national and subnational stakeholders. The commission recognises that the development of implementation rules requires the participation of stakeholders from member states, and it is building a network of spatial data interest communities (SDICs) to assist the drafting teams. SDICs bring together "the human expertise of users, producers and transformers of spatial information, technical competence, financial resources, and policies, with an interest to better use these resources for spatial data management and the development and operation of spatial information services."

New kinds of organisations. A European SDI framework will likely require new kinds of organisations. The simplest case is the merger of various government departments with responsibilities for collecting geographic information. The driving force behind this restructuring is the perceived administrative benefits associated with the creation of an integrated database for the agency as a whole. Alternatively, a special government agency might be set up outside the existing governmental structure with a specific mission to maintain and disseminate core datasets.

In other cases, joint ventures between groups of stakeholders may be necessary. The Large-scale Basemap of the Netherlands described in chapter 3 is a good example of a joint venture to create and maintain key elements of a spatial data infrastructure. This involves a continuing commitment to share the costs involved between a number of public-sector agencies from central and local governments as well as private utility companies. The national joint venture agency that manages the project is a public–private partnership of data users that has been operating at the local, provincial, and national levels for more than 10 years.

KEY STRATEGIC ISSUES

The key issues for SDI development have been previously identified as machinery for coordination, metadata services, capacity building, and data integration (Masser 2001). These four issues will still play a vital role in the success of future SDIs, but they need to be modified in light of recent developments.

Governance. It is necessary to go beyond establishing the mechanics of SDI coordination and give top priority to the creation of appropriate SDI governance structures which are both understood and accepted. This is a daunting task given the number of organisations that are likely to be involved. For example, there are more than 100,000 municipalities in the European Union engaged in SDI-related activities. Obviously it is not always possible to bring all the stakeholders together for decision-making purposes, and structures must be devised for keeping all of them informed and giving them an opportunity to have their opinions heard. The simplest solution to this problem is to create hierarchical structures at the national, state, and local levels. They should be as inclusive as possible from the outset of an SDI initiative, so that all those involved can develop a shared vision and feel a sense of common ownership. This may slow down things in the short term, but building a base for future collaboration is an essential prerequisite for long-term success.

Facilitating access. One of the biggest problems faced by users is lack of metadata. Without appropriate metadata services it is unlikely that SDIs will achieve their overarching objective of maximising the use of geographic information. One very practical reason for giving high priority to metadata services is that they can be developed relatively quickly and cheaply. The development of spatial portals has also opened up new possibilities in that they provide points of entry to SDIs, help users around the world to connect to GI resources, and allow GI users and providers to share content and create consensus. The activities of the Catalonian IDEC initiative described in chapter 3 demonstrate how spatial portals are fostering data sharing in the region and maximising the use of existing data resources. Initiatives such as IDEC can profoundly change existing administrative cultures and promote more widespread commercialisation of information gathered by public-sector organisations.

Building capacity. SDIs succeed when they maximise the availability of local, national, and global GI assets. Organisational-change management plays a key role in SDI development. Capacity building is important in less developed countries, where the implementation of SDI initiatives often depends on a limited number of staff with the necessary GI management skills. But development of GIS capabilities also requires a great deal of work in many industrialised countries, particularly at the local level. The online consultations and the public hearing in Rome discussed in chapter 4 suggest that the INSPIRE initiative does not adequately address capacity building, notwithstanding the commission's efforts to encourage the participation of users in spatial data interest communities.

Interoperability. In countries where large-scale topographic datasets are incomplete, the creation of a national digital topographic database can be an expensive, long-term task. SDI development must therefore exploit alternative information sources such as remote-sensing data and crowd sourcing, in addition to conventional survey technology.

NETWORKING, NETWORKING, NETWORKING

The starting point for this book was the growing importance of location in modern information society. However, as described above, effective networking is key to successful SDIs, and an alternative to the slogan "location, location, location" might be "networking, networking, and (more) networking." This means that SDI stakeholders must be more proactive than in the past in seeking out partners and forming coalitions to facilitate SDI development and implementation. Networking must also be seen as a social learning process in which people from different backgrounds with different mandates have to work alongside each other to find solutions that are acceptable to all. This can be achieved only by understanding and respecting each other's positions. The six practitioners profiled in chapters 3 and 4 come from diverse scientific and technical backgrounds, but networking and data sharing feature prominently in the views of each. As Alessandro Annoni put it (box 4.4),

> INSPIRE is an interesting model for developing not only a technological infrastructure, but also shared practices and working methods, thorough collaboration, and partnership. It takes a lot of time and effort, but it is well worth it to achieve a shared sense of ownership of both process and outcomes.

References

Blakemore, M., M. Craglia, K. Evmorfopolou, A. Fonseca, C. Gouveia, A. Lefevre, I. Masser, and P. Pekkinen. 1999. Comparative evaluation of national spatial data infrastructures. Methods for access to data and metadata in Europe (MADAME) project report, Sheffield: University of Sheffield.

Chan T. O., E. Thomas, and B. Thompson. 2005. Beyond SDI: The case of Victoria. *International Archives of Photogrammetry, Remote Sensing and Spatial Information Sciences,* 36 (4/W6): 47–52.

Commission of the European Communities (CEC). 1985. On the adoption of the Commission work programme concerning an experimental project for gathering, coordinating and ensuring the consistency of information on the state of the environment and natural resources in the Community, Council Decision 85/338.EEC. *Official Journal of the European Commission* L176.

——. 2000a. *European Union enlargement: A historic opportunity.* Brussels: DG Enlargement.

——. 2000b. A framework for Community action in the field of water policy, Directive 2000/60/EC of the European Parliament and of the Council. *Official Journal of the European Communities* L327.

——. 2001a. *ESDI organisation and E-ESDI action plan.* Brussels: Environment Directorate.

——. 2001b. Draft report of the Vienna meeting of 17th December 2001. Brussels: Environment Directorate.

——. 2001c. *EU focus on coastal zones: Turning the tide for Europe's coastal zones.* Luxembourg: Office for Official publications of the CEC.

——. 2002a. Decision 1600/2002/EC of the European Parliament and of the Council of 22 July 2002 laying down the Sixth Community Environment Action programme. *Official Journal of the European Union* L242.

——. 2002b. Memorandum of understanding between Commissioners Wallstrom, Solbes and Busquin: Infrastructure for spatial information in Europe (INSPIRE). Brussels: European Commission. http://inspire.jrc.ec.europa.eu/.

——. 2003. The re-use of public sector information. Directive 2003/98/EC of the European Parliament and of the Council. *Official Journal of the European Union* L345: 90–96.

——. 2004. Proposal for a Directive of the European Parliament and the Council establishing an infrastructure for spatial information in the Community (INSPIRE). COM (2004) 516 final. Brussels: Commission of the European Communities.

——. 2006. Communication from the Commission to the European Parliament pursuant to the second subparagraph of Article 251 (2) of the EC Treaty concerning the common position of the Council on the adoption of a Directive of the European Parliament and of the Council establishing an infrastructure for spatial information in the Community (INSPIRE). COM (2006) 51 final. Brussels: Commission of the European Communities.

——. 2007a. Directive 2007/60/EC of the European Parliament and the Council of 23 October 2007 on the assessment and management of flood risks. *Official Journal of the European Union* L288: 28–34.

——. 2007b. Directive 2007/2/EC of the European Parliament and the Council of 14 March 2007 establishing an infrastructure for spatial information in the Community (INSPIRE). *Official Journal of the European Union* L326:12–30.

——. 2008a. Commission regulation (EC) No 1205/2008 of 3 December 2008 implementing Directive 2007/2/EC of the European Parliament and of the Council as regards metadata. *Official Journal of the European Union* L326: 12–30.

——. 2008b. Towards a shared environmental information system (SEIS). COM (2008) 46 final. Brussels: Commission of the European Communities.

——. 2009a. Commission decision of 5 June 2009 implementing Directive 2007/2/EC of the European Parliament and of the Council as regards monitoring and reporting. *Official Journal of the European Union* L148: 18–26.

——. 2009b. Commission regulation (EC) No 976/2009 of 19 October 2009 implementing Directive 2007/2/EC of the European Parliament and of the Council as regards the network services. *Official Journal of the European Union* L274: 9–18.

Craglia, M., A. Annoni, M. Klopfer, C. Corbin, L. Hecht, G. Pichler, and P. Smits, eds. 2003. *Geographic information in the wider Europe.* Sheffield: University of Sheffield. www.ec-gis.org/ginie (last accessed 28 October 2009).

Craglia, M., and M. Campagna, eds. 2009. Advanced regional spatial data infrastructures in Europe. JRC Scientific and Technical Reports. Luxembourg: Office for Official Publications of the European Communities.

Craglia, M., K. Fullerton, and A. Annoni. 2005. INSPIRE: An example of participative policy making in Europe. *GEOinformatics* Sept. 8: 43–7.

Crompvoets, J., A. Rajabifard, A. Bregt, and I. Williamson. 2004. Assessing the world wide developments of national spatial data clearinghouses. *International Journal of GIS* 18: 1–25.

Dangermond, J. 2006. How will we use spatial information in the future and how can a common EU infrastructure for such information contribute to development. Presentation at the EU Interparliamentary conference on the INSPIRE Directive. April 2–4 2006, Gavle, Sweden.

Department of the Environment. 1987. Handling geographic information: Report of the Committee of Enquiry Chaired by Lord Chorley. London: HMSO.

Department of Finance and Personnel (DFP). 2009. *Northern Ireland Geographic Information Strategy 2009–2019: Effectively using information of geographic location.* Belfast: Department of Finance and Personnel.

Department of Sustainability and Environment (DSE). 2005. *Victorian Spatial Information Strategy 2004–2007.* Melbourne: Department of Sustainability and Environment. www.land.vic.gov.au.

Executive Office of the President. 1994. Coordinating geographic data acquisition and access, the National Spatial Data Infrastructure, Executive Order 12906. *Federal Register* 59: 17671–17674.

Finnish National Council for Geographic Information. 2004. National geographic information strategy 2005–2010. Ministry of Agriculture and Forestry publication 10a, Helsinki, Finland.

Garcia Almirall, P., M. Moix Bergada, and P. Queralto Ros. 2008. The socio-economic impact of the spatial data infrastructure of Catalonia. JRC Scientific and Technical Reports, Luxembourg: Office for Official Publications of the European Communities.

GeoConnections. 2005. *The Canadian Geospatial Data Infrastructure: Roadmap achieving the vision of the CGDI.* Ottawa: GeoConnections. www.geoconnections.org

Guimet, J. 2006. From theory to practice: The SDI platform and its application to public and private sectors in Catalonia. Proc. INSPIRE workshop, Ispra, Italy.

INSPIRE. 2003. Contribution to the extended impact assessment of INSPIRE. Ispra, Italy: Joint Research Centre. www.ec-gis.org/inspire/.

——. 2004. INSPIRE work programme preparatory phase 2005–6. Ispra, Italy: Joint Research Centre. www.ec-gis.org/inspire/.

International Commission for the Protection of the Danube River. 2005. The Danube river basin district: River basin characteristics, impact of human activities, and economic analysis required under Article 5, Annex II and Annex III and inventory of protected areas required under Article 6, Annex IV of the Water Framework Directive (2000/60/EC). Vienna: International Commission for the Protection of the Danube River.

Lenk, M., A. von Domming, and R. Mordhorst. 2008. Implementation of SDI in Germany: GDI-DE technical architecture and organisation model. Proc 2nd INSPIRE conference. Maribor, Slovenia. http://inspire.jrc.ec.europa.eu/.

Longley, P., M. Goodchild, D. Maguire, and D. Rhind, 2005. *Geographic information systems and science.* 2nd ed. Chichester: John Wiley and Sons.

Lucchi, R., and M. Millot. 2009. INSPIRE invoke services: Survey of requirements, challenges and recent research findings supporting the development of invoke spatial data specification. JRC Scientific and Technical Reports, Luxembourg: Office for Official Publications of the European Communities.

Masser, I. 2001. The Indian National Geospatial Data Infrastructure. *GIM International* 15 (8): 37–9.

——. 2005. *GIS worlds: Creating spatial data infrastructures.* Redlands, Calif.: ESRI Press.

——. 2006. Multi level implementation of SDIs: Emerging trends and key strategic issues. *GIM International* 20 (2): 31–3.

Masser, I., and J. Rix. 2009. Implementing SDIs: The subnational dimension. *GeoInformatics* (December 15). www.geoinformatics.com.

Murre, L. 2002. GBKN: The large scale base map of the Netherlands. Proc. GSDI 6. Budapest, Hungary.

National Research Council. 1993. *Toward a coordinated spatial data infrastructure for the nation.* Mapping Science Committee. Washington, D.C.: National Academy Press.

Nebert, D. D., ed. 2004. *Developing spatial data infrastructures: The SDI cookbook.* Version 2.0. Reston, Va.: FGDC. www.gsdi.org.

Nemoforum. 2001. *The National Geoinformation Infrastructure of the Czech Republic: Programme for the years 2001–2005.* Prague, Czech Republic: Nemoforum.

Ordnance Survey Northern Ireland (OSNI). 2003. *A geographic information strategy for Northern Ireland: A consultation document.* Belfast: Ordnance Survey Northern Ireland.

Peersmann, M., H. van Eekelen, and M. Meier. 2009. The large scale topographic map of the Netherlands (GBKN): The transition from a public private partnership (PPP) to a legally mandated key registry (BGT). GSDI 11, Workshop 4.5. www.gsdi.org.

PIRA International Ltd, University of East Anglia, and Knowledge Ltd. 2000. *Commercial Exploitation of Europe's Public Sector Information.* Luxembourg: EC DG INFSO.

Rajabifard, A., A. Binns, I. Masser, and I. Williamson. 2006. The role of sub national government and the private sector in future spatial data infrastructures. *International Journal of Geographical Information Science* 20: 727–741.

Rajabifard, A., M. E. Feeney, I. Williamson, and I. Masser. 2003. National spatial data infrastructures. In *Development of spatial data infrastructures: From concept to reality,* eds. I. Williamson, A. Rajabifard, and M. E. Feeney. London: Taylor and Francis.

Riecken, J., and L. Bernard. 2003. North Rhine Westphalia: Building up a regional SDI in a cross border environment. Proc. 9th EC-GI and GIS workshop. La Coruna, Spain

Salge, F., E. Ladurelle-Tikry, L. Fourcin, and B. Dewynter. 2009. Review of sub national SDIs in France: An outcome of the eSDI-Net+ project. Proc. GSDI 11. www.gsdi.org.

Steenson, T. 2009. Spatially enabling Northern Ireland. Proc AGI Conference. London: Association for Geographic Information.

United Nations Economic Commission for Europe (UNECE). 1998. Convention on access to information, public participation in decision making and access to justice in environmental matters (The Aarhus Convention). Geneva: United Nations Economic Commission for Europe.

Urbanas, S. 2008. Stepwise approach developing a Lithuanian geographic information infrastructure. Proc. 2nd INSPIRE Conference, Maribor. http://inspire.jrc.ec.europa.eu/.

Vandenbroucke, D., and K. Janssen. 2008. Spatial data infrastructures in Europe: State of play spring 2007. Summary report of a study commissioned by the EC (EUROSTAT) in the framework of the INSPIRE initiative, Spatial Applications Division, Katholieke Universitiet Leuven (SADL). http://inspire.jrc.ec.europa.eu/.

Van Orshoven, J., P. Beusen, M. Hall, C. Bamps, D. Vandenbroucke, and K. Janssen, 2003. Spatial data infrastructures in Europe: State of play spring 2003. Summary report of a study commissioned by the EC (EUROSTAT and DGENV) in the framework of the INSPIRE initiative, Spatial Applications Division, Katholieke Universitiet Leuven (SADL). http://inspire.jrc.ec.europa.eu/.

Villa, M., G. Di Matteo, R. Lucchi, M. Millot, and I. Kanellopolous. 2008. *SOAP primer for INSPIRE discovery and view services.* Luxembourg: Office for Official Publications of the European Communities. http://inspire.jrc.ec.europa.eu/.

Weissbord, M., and S. Janoff. 2000. *Future Search: An action guide to finding common ground in organisations and communities.* 2nd ed. San Francisco: Berrett-Koehler.

Zimova, R. 2009. *10 years of Nemoforum, 1999–2009.* Prague, Czech Republic: Nemoforum.

Abbreviations

ACTIG	Catalan Association of GI Technologies
AESIG	Spanish national geographic information association (Asociacion Espanola de Sistemas de Informacion Geografica)
AFIGéO	Association Française pour l'Information Géographique
AGI	Association for Geographic Information, United Kingdom
AGILE	Association of Geographic Information Laboratories in Europe
AGINI	Association for Geographic Information Northern Ireland
AM/FM	Automated mapping and facility management
ANZLIC	Australia New Zealand Land Information Council
ASDI	Australian Spatial Data Infrastructure
BGT	Large-scale Topography of the Netherlands (Basisregistratie Grootschalige Topografie)
BKG	German Federal Organisation for Cartography and Geodesy
CAGI	Czech Association for Geographic Information
CASA	Centre for Advanced Spatial Analysis, University College London
CEC	Commission of the European Communities
CENIA	Czech Environmental Information Agency
CEODE	Centre for Earth Observation and Digital Earth, China
CGDI	Canadian Geospatial Data Infrastructure
CNIG	French National Council for Geographic Information (Conseil National de l'Information Géographique)
COGI	Committee on Geographic Information, European Commission
CORINE	Coordination of Information on the Environment programme
DBMS	Database management system
DCAL	Department of Culture Arts and Leisure, Northern Ireland
DDGI	German national geographic information association
DFP	Department of Finance and Personnel, Northern Ireland
DG	Directorate General, European Commission
DGH	Directorate General of Hydrocarbons, India
DSE	Department of Sustainability and Environment, Victoria, Australia
EC	European Commission
E-ESDI	Environmental European Spatial Data Infrastructure initiative (precursor to INSPIRE)
EIA	Environmental Impact Analysis
ESA	European Space Agency
EU	European Union
EuroGeographics	Association of European mapping, land registry, and cadastral agencies
EuroGeoSurveys	European Geological Surveys Association
EUROGI	European Umbrella Organisation for Geographic Information
FGDC	Federal Geographic Data Committee, United States
FIG	International Federation of Surveyors

FIS	Fellow of the Irish Institution of Surveying
FIsntCES	Fellow of the Institution of Civil Engineering Surveyors
GALILEO	EU satellite navigation project
GBKN	Large-scale Basemap of the Netherlands (Grootschalige Basis Kaart Nederland)
GDI	Geospatial Data Infrastructure
GDI-DE	German Spatial Data Infrastructure
GEO	Group on Earth Observations, European Commission
Geonovum	Dutch National Geographic Information Association and Clearing House
GEOSS	Global Earth Observation System of Systems
GI	Geographic information
GII	Geographic Information Infrastructure, Lithuania
GI 2000	European Commission initiative
GINIE	Geographic Information Networks in Europe
GIP	Dutch Subsurface Data and Information Programme
GIS	Geographic information system(s)
GISCO	GIS for the Commission
GMES	Global Monitoring of Environment and Security initiative
GOS	Geospatial One Stop, United States
GPS	Global Positioning System
GSDI	Global Spatial Data Infrastructure Association
ICC	Catalunyan Institute for Cartography
IDEC	Catalunyan SDI (Infraestructura Dades Especials de Catalunya)
IDEE	Spanish Spatial Data Infrastructure
IGN	French national mapping agency (Institut Géographique National)
IMAGI	German Federal Interministerial Committee for Geoinformation (Inter Ministerieller Auschuss fur Geoinformationswesen)
INSPIRE	Infrastructure for Spatial Information in Europe
INTERREG	European Community initiative concerning border development and cross border cooperation
IPR	Intellectual property rights
IR	Implementing rules (INSPIRE)
ISDE	International Society for Digital Earth
ISDI	Irish Spatial Data Infrastructure
ISO	International Standards Organisation
JRC	Joint Research Centre, European Commission
KPN	Dutch Telecom
LGA	Local government agency
LMO	Legally mandated organisation
LPS	Land and Property Services, Northern Ireland
LSV	National joint venture agency (Landelijk Samenwerkingsverband)
MADAME	Methods for Access to Data and Metadata in Europe

MARS Monitoring Agriculture with Remote Sensing project, JRC
MCIM Member of the Chartered Institute of Marketing
MOU Memorandum of understanding
MSC Mapping Sciences Committee, National Research Council, United States
NDR National Data Repositories Work Group
Nemoforum Czech National Platform for GI Strategy Coordination
NGO Nongovernmental organization
NI Northern Ireland
NIMA Northern Ireland Mapping Agreement
NITG-TNO National Institute for Applied Geoscience of the Netherlands
NLS National Land Survey
NSDI National spatial data infrastructure(s)
OGC Open Geospatial Systems Consortium
OSNI Ordnance Survey Northern Ireland
PHARE EU Pologne, Hongrie Assistance à la Reconstruction Economique programme
PPP Public–private partnership
PSI Public-sector information
PSPP Public Services Productivity Panel, UK
R & D Research and development
RICS Royal Institution of Chartered Surveyors, UK
RGE French large-scale map project (Referentiel Géographique a grand Echelle)
RGI Users Advisory Council of the Space for Information
SADL Spatial Applications Division of the Katholieke Universiteit Leuven, Belgium
SAR Synthetic aperture radar
SARS Severe acute respiratory syndrome
SDI Spatial data infrastructure(s)
SDIC Spatial data interest community
SEIS Shared Environmental Information System, European Commission
SOAP Simple object access protocol
UELN United European Levelling Net
UK United Kingdom
UKLC United Kingdom Location Council
UNECE United Nations Economic Commission for Europe
URISA Urban and Regional Information Systems Association, United States
USGS United States Geological Survey
VNG Association of Dutch Local Authorities
VROM Ministry of Housing, Spatial Planning, and the Environment
VSIS Victorian Spatial Information Strategy, Australia
WCS Web coverage service
WFS Web feature service
WMS Web mapping service

RELATED TITLES FROM ESRI PRESS

GIS Worlds: Creating Spatial Data Infrastructures

ISBN: 978-1-58948-122-0

GIS Worlds: Creating Spatial Data Infrastructures discusses the evolution of SDIs around the world, shows where SDIs are advancing, and describes where more work is needed. More significantly, it details the implementation of SDIs from a practical perspective and outlines a method of institution building for regional, continental, and global SDIs.

Research and Theory in Advancing Spatial Data Infrastructure Concepts

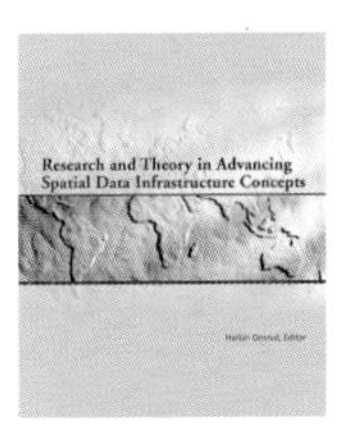

ISBN: 978-1-58948-162-6

Presenting the latest research by renowned international experts, *Research and Theory in Advancing Spatial Data Infrastructure Concepts* analyzes models for planning, financing, and implementing SDI initiatives and assesses the extent to which established SDI projects in Australia, India, and the European Union are contributing to national economic competitiveness and social well-being.

Spatial Portals: Gateways to Geographic Information

ISBN: 978-1-58948-131-2

Spatial Portals: Gateways to Geographic Information examines how spatial portals are revolutionizing the way GIS professionals find, share, and apply knowledge—from the local level to the world stage. Spatial portals help people search for and access networks of relevant information held by governments and other organizations. In so doing, portals help people quickly evaluate options and make better decisions that can save time, money, and even lives.

Lining Up Data in ArcGIS: A Guide to Map Projections

ISBN: 978-1-58948-249-4

Lining Up Data in ArcGIS: A Guide to Map Projections is an easy-to-navigate, troubleshooting reference for any GIS user with the common problem of data misalignment. Complete with full-color maps and diagrams, this book presents techniques to identify data projections and create custom projections to align data. Formatted for practical use, each chapter can stand alone to address specific issues related to working with coordinate systems.

ESRI Press publishes books about the science, application, and technology of GIS. Ask for these titles at your local bookstore or order by calling 1-800-447-9778. You can also read book descriptions, read reviews, and shop online at www.esri.com/esripress. Outside the United States, contact your local ESRI distributor.